Crickets and Household Pests

Crickets and Household Pests

Professor T.V. Sathe
Entomology Division
Department of Zoology
Shivaji University
Kolhapur – 416 004
&
Mrs. Manisha R. Awate
Department of Zoology
Shivaji University
Kolhapur – 416 004

2009
DAYA PUBLISHING HOUSE
Delhi - 110 035

ISBN 81-7035-610-5
ISBN 978-81-7035-610-3

Published by : **Daya Publishing House**
1123/74, Deva Ram Park
Tri Nagar, Delhi - 110 035
Phone: 27383999
Fax: (011) 23260116
e-mail : dayabooks@vsnl.com
website : www.dayabooks.com

Showroom : 4760-61/23, Ansari Road, Darya Ganj,
New Delhi - 110 002
Phone: 23245578, 23244987

Laser Typesetting : **Classic Computer Services**
Delhi - 110 035

Printed at : **Chawla Offset Printers**
Delhi - 110 052

PRINTED IN INDIA

Preface

Household pests are associated with human dwellings, prisons, barns, godowns, libraries etc. Many of them are cosmopolitan in distribution and transmit several fatal diseases to human being and domestic animals. New insect pest species are being continuously added in the environment, are responsible for additional problems. Hence, their control is the need of the day. The book contains information on international status, distribution, identification, nature of damage and epidemiological role, life cycle, and control measures of 45 household insect pests.

Authors are thankful to Shivaji University authority for providing facilities for completion of this book. We hope that the book will be useful to common people, students, teachers, scientists and consultants in the field of household pest management.

T.V. Sathe

Manisha Awate

Contents

Chapter 1

Cricket Diversity and Control

INTRODUCTION

The dominant group of animals on the earth is insects which surpass all other terrestrial animals in number and occur practically everywhere. Several hundred thousand different kinds have been described, three time as many as there are in the rest of the animal kingdom and some authorities believe that the total number of different kinds may approach 30 million (Gupta, 2004). More than a thousand kinds may occur in fair-sized backyard, their populations often number many millions per acre (Gupta, 2004).

Many insects are extremely valuable to humans. Society could not exist in its present form without them. Insect pollinating activities make possible higher production of many agricultural crops, including many orchard fruits, nuts, clovers, vegetables, cotton and tobacco, etc. Insects provide us honey, bee wax, silk and other products of commercial value. They serve as food for many birds, fishes and other beneficial animals including man (Shrivastava, 1996). They perform valuable services as scavengers and helping to keep harmful animals and plants in check. They are useful in medicine and in scientific research and are looked upon as interesting animals by people in all walks of life. A few insects are harmful and cause enormous losses each year to agricultural, horticultural and forest crops and stored products. They may transmit diseases that seriously affect the health of plants and humans and other animals.

Majority of insects are relatively small, probably three-fourth or more, are less than 6 mm in length. Their small size make them suitable to live in places that would not be possible to larger animals. Insects measure in size from about 0.25 to 330 mm in body length and from about 0.5 to 300 mm in wing span. One fossil dragonfly had a wing spread of over 760 mm. Some of the longest insects are very slender measuring about 330 mm is a walking stick occurring in Borneo but, some beetles have a body nearly as large as one's first (Borror *et al.*, 1982). The largest insects in North America are some of the moths, with a wing span of about 150 mm, and the walking sticks, with a body length of about 150 mm (Gupta, 2004). Insects are the only invertebrates with wings, and these wings have had an evolutionary origin different from that of the vertebrates. Insects have different colours from very drab to brilliant. No other animals on the earth are more brilliantly coloured than some of the insects. Insects are cold-blooded creatures. They have tremendous reproductive capacity. Most people do not realize just how great they are.

Insects have different feeding habits. They feed on an almost variety of foods. They are defoliaters, sap suckers, blood suckers, gall formers, borers, leaf miners, leaf rollers, etc. Thousands of species feed on plants and practically every kind of plant (on land or in fresh water) is fed upon by some kind of insects. The plant feeders may feed on almost any of the plant part; caterpillars, beetles and leaf hoppers feed on the leaves; aphids feed on the stems, leaves, flowers; white grubs feed on the roots; certain weevils and moth larvae feed on the fruits and so on. Some insects burrow into the soil and feed on roots. Thousands of insects are carnivorous, feeding on other animals. Some are predators and some are parasitoids. Many insects that feed on vertebrates are blood sucking. Some of the parasite insects such as mosquitoes, lice, fleas and certain bugs not only are annoying pests because of their bites, but also serve as disease vectors. Some insects feed on dead woods, others feed on stored food of all types. Some feed on various fabrics and many feed on decaying materials.

Crickets are very important group of Class Insecta since they are visualized as pest insects in western and other countries. They are hardy insects and are not damaged by being picked up and handle. They do not bite and do not transmit any diseases. House crickets, field crickets, camel crickets, ant crickets, scaly crickets etc. are some important types of crickets. Crickets are somewhat related to grasshoppers and more closely related to katytids or bush crickets. They have somewhat flattened bodies and long antennae. Crickets belong to the family Gryllidae of Order Orthoptera. According to Vasanth(1993) there are about 2300 species of crickets in the world. Crickets tend to be nocturnal and are often confused with grasshoppers. Crickets are popular pets and are considered goodluck in Asia especially, China where they are kept in cages (Carrera, 1991).

House crickets are 3/4 to 1 inch long, light yellowish brown with three dark brown bands on the head. They are nocturnal and prefer warm areas such as kitchen, basement, fireplace, cracks and crevices, behind base boards, etc. They make distinctive, annoying chirping sound. They eat just about anything but, love bread crumbs and liquids, including beer, often ruin stored clothing.

Field crickets are 1/2 to 1-1/4 inch long and black in body colouration. They found in pastures, meadows, lawns, occasionally indoors. They also eat almost anything but are associated with field crops, especially wheat, oats, rye, alfalfa and several plants, dead insects, seeds, leather, paper and old cloths, etc. They also feed on textiles of cotton, linen, wood and silk. Usually, one generation is completed in a single year. They generally hatch in May and comeout in massive number with swarms covering many square miles. Several species are recognized which differ chiefly in habits, life history and song. Most species of *Gryllus* chirp but one species occurring in the Southwest, *G. rubens* Scudder, produces a more or less long trill. The most common species of *Gryllus* in the East is probably the Northern field cricket, *Gryllus pennsylvanicus* Burmeister.

Crickets can be found in many locations most species can be found in grasslands and forests. Scaly crickets *Pseudomogoplistes scamper* can be found on sea shores. Bod bush crickets *Merioptera brachytera* Linnaeus can be found in logs and marshes. Camel crickets are natural cave dwellers. The mole crickets *Gryllotalpa gryllotalpa* Latreille can be found underground, many other species live in trees, ex. the Oak bush cricket *Meconoma thalassinum* De Geer while, the ant cricket *Mymecophilus acervorum* Panzer lives only in ant nests.

Camel crickets are light tan to dark brown coloured, measuring about 1/2 to 1-1/2 inch in body length. They have hump-back appearance and found in cool, damp areas, such as under logs or stones and in crawl spaces, basements and attics. They are nocturnal and they feed on clothes and lace curtains. Females lay eggs in spring, which hatch around April. Only one generation is completed in a year by this cricket. Unlike other crickets, they do not chirp. They become immobile in the presence of strong light. Field crickets, in adult stage spend their days in shallow burrows beneath a stone, clod of dirt or a tuft of plant. They are most active during the night. Males begin their nightly serenading to attract female mates. Females do not stridulate or rub special body structures together for producing sound. At the front of each wing male has a heavy vein with a row of teeth on the under side. According to some workers, the pitch of the chirps is slightly higher than the highest octave on a piano. The warmer the night, the faster they chirp. Crickets have their special songs for courtship, fight and sound.

China is very famous country in the World for organizing cricket fightings and songs. Cricket fightings were very popular in ancient China. The practice became rare after the revolution, due to its 'Bourgeois nature'. Now it is making a come back. In Beijing there are many associations for organizing crickets. The associations sponsor national tournaments whereby modern equipment such as video cameras are used to zoom in and project the fighting onto many television sets which enable many viewers to see the fighting simultaneously.

The female of field cricket lays her eggs in the fall, approximately two weeks after maturation and development of wings. Firm base soil sites are preferred for egg laying. A single female lays 150 to 400 eggs. Eggs remain in the soil throughout the winter and hatch the following spring. Thus, eggs hatched into nymphs. Nymphs can be identified by the incomplete development of the wings. Nymphal period lasts for about three months. Nymphs become adults. The adult stage is capable for fly and mate.

Cricket outbreaks are also common when large number of nymphs complete their development and embark on night time mating flights. The largest cricket outbreaks seem to occur during days of springs and summers. The reason for cricket outbreaks under such conditions is not fully understood. Although crickets are locally abundant throughout the year but, numbers appear to be highest in August and September when a summer drought is broken by rainfall and cooler weather.

Field crickets become a considerable household nuisance in abundant condition. They are primarily outdoor insects and as such are on accidental indoor invaders. Unlike house crickets, they do not breed or establish permanently indoor. However, they damage clothing, drapes or wall covering by their feeding activities or by staining with their feces or regurgitations. Crickets can damage fabrics, soiled clothing or clothes stained with perspiration. During severe outbreaks crickets can create an aesthetically unacceptable situation around places of business. Dead crickets quickly pile up, causing odors and become repulsive to many people.

Crickets get attracted to bright light. Buildings that are brightly light at night are most likely to attract the largest number of crickets. Reducing outdoor lights is the first and most important step in a cricket control programme. Low pressure sodium vapour lamps and yellow incandescent "bug lights" are less attractive to crickets than standard incandescent, fluorescent, mercury vapour or halogen lights. Lighting schedule should be restricted to a few hours each night for reducing cricket populations.

Crickets are especially likely to enter cracks and openings around outdoor lights. They should be checked carefully. Steel or brass wool in weep holes are temporary insect barriers.

Crickets around building perimeters can be killed with any of several insecticidal baits and sprays. Baits for crickets include product containing hydramethylnon (MaxforceÒ and CombatÒ granular, ant and cricket baits), metaldehyde, carbaryl (Green Light Bug and Snail Bait) and trichlorform (Larva-Lur™).

Liquid insecticide sprays may be applied to outdoor sites around weep holes, doorways, windows and other cricket entryways. Indoor sprays are not very effective for field crickets and are not recommended. Pesticides found effective against crickets refers to Chlorpyrifos (DursbanÒ, Carbaryl (SevinÒ, Permethrin (Spectracic Bug StopÒ and others), Cyfluthrin (Bayers Advanced Home™) and bifenthrin (Ortho Bug B GonÒ).

Cricket control with insecticides should be considered as only partial solution to cricket problems. Insectides should be used in combination with reduction in outdoor lighting for best control.

Mole crickets are relatively common but, because they are nocturnal and spend nearly all their lives underground in extensive tunnel systems, they are rarely seen. Agricultural fields, lawns and golf courses, etc are the common places where mole crickets inhibit. Except Antarctica they are present in every continent and are commonly considered pests. Very interestingly, in East Asia, they are sometimes used as fried food.

Mole crickets are omnivorus insects. They feed on larvae, roots and grasses. Mole crickets can be predated by birds, rats, skunks armadillous, raccoons and foxes, etc. In some places, mole cricket numbers are declining due to soil erosion and habitat destruction. Along the North Carolina coast, over the past ten years, mole crickets have become the number one insect pest of home lawns, golf courses, municipal and commercial properties and sod farms. Mole crickets remain a challenge to effective management in western countries. Their management requires advanced planning and full-scale programme for best results. However, pest status of these crickets has not been widely attempted in India.

Mole crickets have three basic developmental stages: the egg, the nymph or immature and the adult. Overwintering takes place as either in last nymphal stage or adult that can remain somewhat active throughout the winter. In winter they cause damage by short tunnel mounds. As the soil temperature rises in March and April, the crickets do more tunneling. Crickets that over wintered as nymphs finish their development into adults during the spring. Only one generation is completed during a year by the mole cricket.

In late March and early April, the adults begin to fly and mate. Mole cricket prepares chamber and acts much like a megaphone. A soft toad like call is produced by male for about an hour after sunset specially, at warm evenings in April. The call attracts the female for mating. Egg laying takes place shortly after mating at the end of April or early May. Mated female lays her eggs at the depth of 3 to 10 inches in the soil. She usually constructs three to five chambers and lays a total of 100 to 150 eggs. Eggs are rounded, translucent and white. They hatch within 20 days between late May and July. The Southern mole crickets may continue to lay eggs throughout the summer and incubation occurs in August and September. The males die after mating while, female die shortly after completion of egg laying.

The newly emerged nymphs are small, about ¼ inch in length. There are 6 to 8 instars in nymphal stage. Small nymphs do not show wings, but the larger nymphs have small wing pads. Crickets continue to feed and grow through the summer and become most destructive in late August to early October. A mole cricket can prepare a tunnel of a length of 10 feet. Mole cricket activity is regulated by temperature and soil moisture. Mostly, they feed at night. During cold weather, the crickets may stay deep in the soil but, they have their tunneling near the surface in hot weather. Mole crickets damage golf courses, home lawns, recreational areas and sod farms. Damage intensity is increased in the spring and is usually minimal after the middle of May. If the soil is very dry, a second drench of pure water may be necessary to flush the crickets. There is great interest and demand to develop nonchemical control technology for mole crickets in turfgrass. Great advances have been made in nonchemical control in the past few years (Castner, 1987). Biological control of mole cricket has been launched in North Carolina (Castner, 1988).

REVIEW OF LITERATURE

The perusal of literature showed that the crickets have been widely studied with various aspects from different parts of the World (Brunner and Wattenwyl, 1873, 1893; Castner, 1988; Gupta, 1988; Nguyen and Smart, 1992a, 1992b; Frank, 1994). However, from India very little attention is paid on crickets. Mating behaviour of Gryllid crickets have been attempted by Khalifa (1950), Alexander (1957), Alexander and Otte (1967), Cade (1980), Boake (1983), Searcy and Anderson (1986), Boucias *et al.* (1987) Vasanth (1988), Cade and Cade (1992), Mark and David (2001), etc. Nematode parasites of crickets have been studied by Nguyen and Smart (1992a, 1992b), Parkman and Frank (1993) and Parkman *et al.* (1994), etc. Palynology of crickets has been studied by Sailer *et al.* (1984), David (1994), and Frank and Robert (1994). Biology of Crickets has been reported by Fritz (1983), Castner (1987, 1988) and Hudson (1994), etc. Monitoring of crickets have been attempted by Walker (1966, 1982, 1986, 1989, 1996), Short and Koehler (1979) and Forrest (1980, 1983). Migratory behaviour has been studied by Walker and Fritz (1983). Biological control of mole crickets have been studied by Smart and Khuong (1986) and Hudson *et al.* (1988).

Gynandromorphological studies in crickets has been attempted by Omachi (1926), Suzuki (1934), Tweedie (1965), Johnstone (1975), Castner (1988), Frank (1990, 1994), Meagher and Frank (1998), Frank and Parkman (1999) and Frank *et al.* (1995), etc.

The review of literature indicates that Linnaeus (1758) for the first time worked on taxonomy of crickets. Later, Pallas (1771), De Geer (1773), Fabricius (1775), Latreille (1804, 1829), Stoll (1813), Berthold (1827), Serville (1831, 1839), Burmeister (1838), Rambur (1839), Le Guillou (1841), Haan (1842), Guerin-Meneville (1845-1850), Walker (1859, 1869, 1871), Scudder (1869), Giebel (1870), Brunner and Wattenwyl (1873, 1893), Saussure (1874, 1877, 1878, 1897), Pantel (1896), Rehn (1902, 1903), Jacobson and Bianchi (1905), Kirby (1906), Karny (1909), Allard (1910, 1929), Schimmer (1911), Rehn and Hebard (1915), Chopard (1924a, 1924c, 1925a, 1925b, 1925c, 1925d, 1927, 1928b, 1929a, 1929b, 1930, 1931a, 1931b, 1933b, 1936, 1941, 1951, 1954, 1961, 1970), Hebard (1926), Caudell (1927), Shiraki (1930), Fulton (1930, 1952), Uvarov (1935, 1940), Snodgrass (1935, 1937), Khan (1954), Sandrasagara (1954), Fernando (1957a, 1957b, 1957c), Latif and Asghar (1957), Ghouri and Mc Farlane (1958), West and Alexander (1963), Randell (1964), Walker (1966), Bey-Bienko (1966, 1968), Chopard and Dreux (1966), Chopard and Beier (1967, 1968), Mayr (1969), Rao and Mani (1974), Vickery (1977), Townsend (1980), Otte and Cade (1984), etc attempted taxonomy and biodiversity of crickets from various parts of the world.

Scudder (1862, 1869, 1901, 1902) studied the North American Orthoptera. Walker (1869-1871) made a catalogue of crickets containing descriptions of many "new genera" and innumerable "new species" with critical remarks and affinities of the crickets. Brunner and Wattenwyl (1893) have described many new genera and species mostly from Myanmar, Java and Borneo. Rehn (1902, 1903) studied the Orthoptera of New Mexico and western Texas. Some New England Orthoptera have been studied by Allard (1910, 1929). One of the most notable Gryllid-taxonomists, Chopard has not only made extensive studies on African and European crickets, but also published (1918-1970) a great deal of his studies on Gryllids from India, Pakistan, Bangladesh, Sri Lanka, Myanmar, Thailand, Laos, the Malaya Archipelago, Indonesia, Mentawai Islands (Near

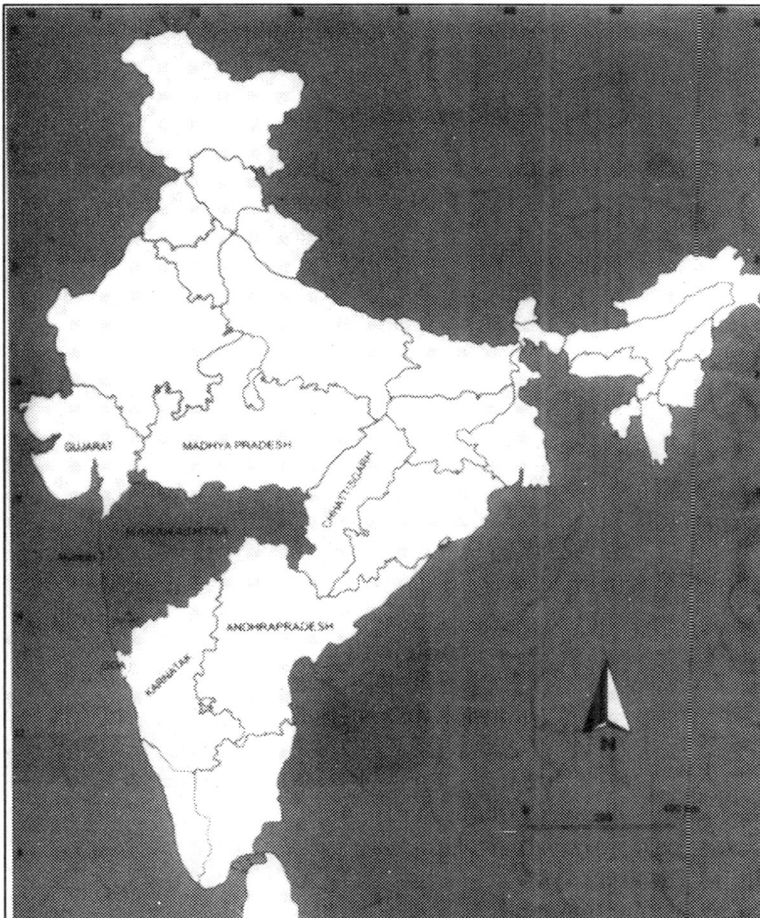

Figure 1: Map of India Showing Maharashtra

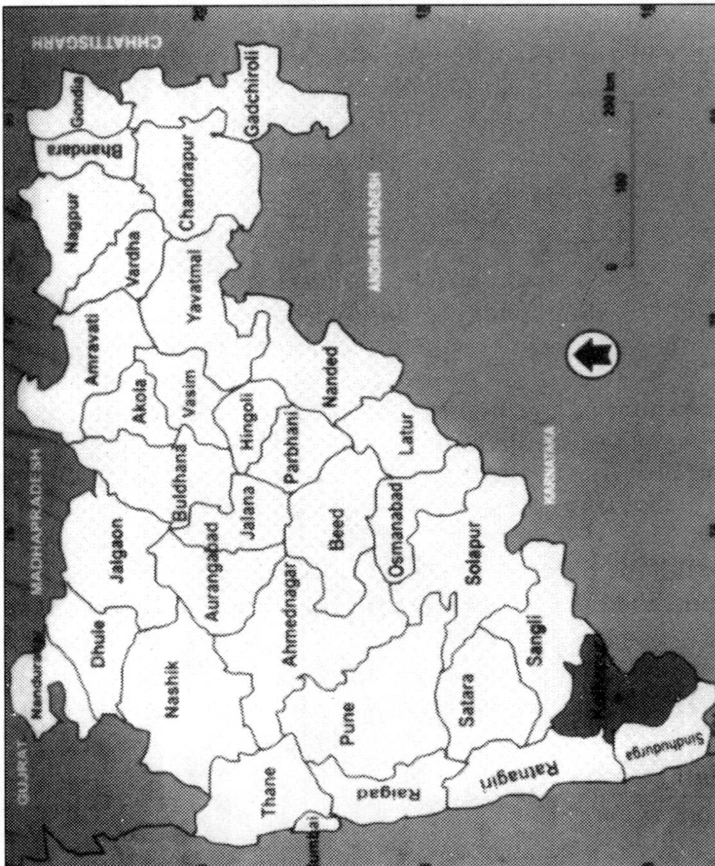

Figure 2: Map of Maharashtra Showing Kolhapur District

Sumtra), New Guinea and the Philippines. Chopard (1925b, 1951) also described and revised the Australian crickets. His contribution on crickets (1954, 1961, 1969) is highly appreciated by Scientists. A new species of *Nemobius* (Orthoptera: Gryllidae) has been reported from North Carolina by Fulton (1930, 1952). The notable work of Sandrasagara (1954) is on the checklist of the Sri Lankan Gryllidae and Bey-Bienko (1966, 1968) on description of many Indonesian and Nepalian species. Chopard and Dreux (1966) described several new species from Nepal.

The review of literature indicates that taxonomy of Indian crickets have been studied by Bolivar (1890, 1900, 1901, 1912, 1922), Anandale (1906), Ghosh (1912), Chopard (1918, 1924b, 1928a, 1929a, 1933a, 1935), Uvarov (1925), Guerin-Meneville (1833), Jones and Chopard (1936), Chopard and Chatterjee (1937), Banerjee and Chatterjee (1955), Chatterjee (1955), Bhowmik (1967a, 1967b, 1967c, 1968, 1969, 1970, 1971, 1975, 1976, 1977a, 1977b, 1977c, 1985), Tandon and Shishodia (1972, 1974), Biswas *et al.* (1975), Vasanth *et al.* (1975), Shishodia and Tandon (1975a, 1975b, 1977), Tandon *et al.* (1976), Bhowmik and Halder (1976), Bhargava (1976, 1981), Sinha and Agrawal (1977), Vasanth (1978, 1980), Kaltenbach (1979), Awate and Sathe (2007, 2008), etc.

Walker (1869-1871) for the first time made records of Gryllids from India. In an earlier work, Saussure (1874) has ignored Walker's British museum catalogue as being of no use unless the types are referred. Saussure (1877-1878) erected the chief pillar in the foundation of the classification of Gryllids. The earliest and the strongest base to Gryllid classification was given by Saussure (1877-1878) who divided the family into 6 tribes Gryllotalpiens, Grylliens, Myrmecophiliens, Oecanthiens, Trigonidiens and Enoptriens based on tarsal form and armature of posterior tibae. He divided each of these, except the Trigonidiens, into legions consisting of numerous genera. Since then there appears to be no general agreement on the taxonomic rank of many groups. Saussure (1894-1897) also modified and elaborated his monographs (1877-1878). Bolivar's (1900) study of South Indian crickets is regarded as the first sincere attempt at a taxonomic study of Indian crickets. Blatchley (1903) studied the Orthoptera (crickets) of India. Kirby (1906) has listed over 80 species from India. Chopard (1924b) was the first to study crickets of Northeast India. He described 3 new genera, each with a new species. Chopard (1935) has recorded 18 species from Meghalaya and Assam including 1 new genus and 4 new species. Jones and Chopard (1936) described a new Gryllid genus from India.

Chopard (1951, 1969) has erected the subfamilies to family status under superfamily Grylloidea.The family Gryllidae has been divided into 10 subfamilies *viz.;* Mogoplistinae, Myrmecophilinae, Scleropterinae, Cachoplistinae, Pteroplistinae, Pentacentrinae, Phalangopsinae, Trigonidiinae, Eneopterinae and Oecanthinae by Chopard and Beier (1967-1968). Many of these have been further subdivided into tribes. Subfamily Oecanthinae has been given family rank, *viz.;* Oecanthidae. Randell (1964) has given a key to the tribes, subtribes and genera of subfamily Gryllinae, founded totally on the characters of the male genitalia. From Northeast India, Randell (1964) in a note worthy work, erected 5 new genera including 2 found in one of these, *viz.; Velarifictorus* has now come to include several species placed in *Scapsipedus* Saussure. The other genera *Plebeiogryllus,* contains only one species, *P. uttiventris* Walker. Walker (1966), in annotated Checklist of the Oecanthinae of the World, gave subfamily rank to Oecanthinae.

Bhowmik (1967-1977) has recorded 14 species of Gryllids from Northeast India. Of these, 3 species of the genera *Anaxipha* Saussure, *Calyptotrypus* Saussure and *Velarifictorus* Randell (described in *Scapsipedus* Saussure) and variety of *Loxoblemmus equestris* Saussure, *viz.; manipurensis,* were described as new. Bhowmik (1969) has placed family Gryllidae under what he calls suborder Grylloidea, which undoubtedly refers to super family Grylloides of most other authors. Chopard's system was broadly followed by Kevan and his co-workers uprating most division in the Orthoptera by one grade. The same system was summerised by Vickery (1977).

A considerable amount of work on the taxonomy of Indian crickets, including the Northeast Indian region have been came in existence from the year 1967. From Arunachal Pradesh and the adjoining areas of Assam in Northeast India, Tandon and Shishodia (1972) have recorded 20 species of Gryllids scattered in 10 genera. Shishodia and Tandon (1975b) have recorded 8 genera from Tripura. In preliminary study on the Gryllidae of Meghalaya Biswas *et al.* (1975) recorded 15 species as new to the state. From Meghalaya city Vasanth *et al.* (1975) described 2 new species of *Velarifictorus* and one of *Teleogryllus*. Vasanth (1978) described for the first time, the females *Velarifictorus jaintianus* Biswas and Ghosh, *V. khasiensis* Vasanth and Ghosh (both described as new by Vasanth *et al.*, 1975) and of *Stephoblemmus humbertiellus* Saussure, all from Northeast India. Biswas *et al.* (1975) and Bhowmik (1977a, 1977b, 1977c) have allowed the system of dividing family Gryllidae into subfamilies while, Tandon and Shishodia (1972) have given family rank to the subfamilies.

Townsend (1980) reported 2 species of *Teleogryllus* Chopard from Northeast India. While, Vasanth (1993) divided the family Gryllidae into 14 subfamilies, *viz.*; Gryllinae, Nemobiinae, Mogoplistinae, Myrmecophlinae, Scleropterinae, Cachoplistinae, Pteroplistinae, Pentacentrinae, Phalangopsinae, Trigonidiinae, Eneopterinae, Podoscirtinae, Itarinae and Oecanthinae. Recently, Vasanth (1993) described a total of 81 species belonging to 39 genera scattered in 12 subfamilies of the family Gryllidae from Northeast India.

Awate and Sathe (2007, 2008) described two new species of the genus *Gryllotalpa* Latreille, namely *Gryllotalpa kolhapurensis* and *Gryllotalpa westerni* from India.

In the present study two families namely, Gryllotalpidae and Gryllidae were selected for taxonomical studies of crickets. For description of species, classification system of Chopard (1933) has been adopted.

The family Gryllotalpidae is characterized by

1. Elongate body, head small and conical.
2. Ocelli two.
3. Pronotum shape convex.
4. Fore leg with coxa very short, femur somewhat curved; tibia very short, triangular, its apical spur transformed into strong, cutting dactyli.
5. There is no mirror and the disk of the tegmina occupied by a large triangular cell.
6. In female ovipositor is completely aborted.

The family Gryllotalpidae is represented by two genera, *Scapteriscus* Scudder and *Gryllotalpa* Latreille.

The family Gryllidae is characterized by following features.

1. Head big and globular.
2. Tarsi compressed.
3. Hind tibiae armed with spines but never serrulated between these spines.

The family Gryllidae is subdivided into two subfamilies *viz.*, Gryllinae and Nemobiinae. The subfamily Gryllinae is characterized by spines immovable on hind tibia and without hairs. The subfamily Gryllinae contains 23 genera namely, *Landreva, Ectolandrevus, Landrevus, Gryllus, Brachytrupes, Acheta, Liogryllus, Gymnogryllus, Scapsipedus, Velarifictorus, Loxoblemmus, Platyblemmus, Stephoblemmus, Parasciobia, Gryllodes, Gryllopsis, Turanogryllus, Cophogryllus, Modicogryllus, Lenigryllus, Plebeiogryllus, Platygryllus, Teleoqryllus.*

In the present work 4 new species of *Gryllotalpa* and 5 new species of *Gryllus* have been described.

MATERIAL AND METHODS

Materials

1) Glass Troughs (Figure 3)

The glass troughs of size 9 × 25 cm and size 10 × 25 cm (height and diameter) were used for rearing of crickets. During the process, the glass troughs were covered with muslin cloth.

2) Test Tubes (Figure 4)

The test tubes of size 15 × 2.5 cm (length and diameter) were used for handling the nymphs and adults of cricket species.

3) Insect Net (Figure 5)

Insect hand net made up of aluminium handle 35 cm long with circular iron ring of 12 cm diameter and ordinary mosquito bag of 35 cm in depth was used for collecting nymphs and adults of both crickets *i.e.* mole cricket and house cricket.

4) Plastic Containers (Figure 6)

Plastic containers (Figure 6) of capacity 1 litre and 1.5 litre were used for keeping/collecting the samples of nymphs/adults of crickets from the natural habitat.

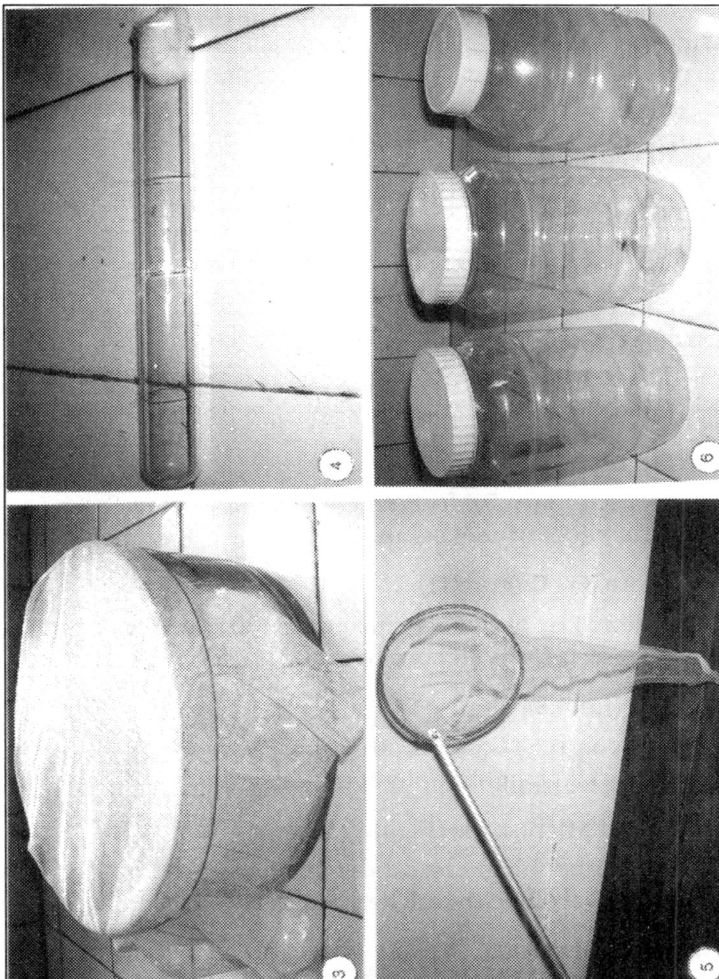

Plate 1

Figure 3: Glass Trough; Figure 4: Test Tubes; Figure 5: Insect Net: Figure 6: Plastic Containers

5) Cricket Rearing Cage (Figure 7)

The cricket rearing box (panpuda local name) made up of steel of size, length 16.5 cm × width 11.5 cm × height 6.5 cm has been used for rearing the crickets and keeping crickets live for longer period for the taxonomical study and photography. The sides of the box were ventilated with small open windows of size, length 20.00 mm × width 3.00 mm. 6 windows were into the width side and 12 into the length side.

6) Slide Boxes (Figure 8)

Slide boxes were used for keeping the permanent slides safely. Slide boxes of size, 28 × 22 × 3.5 cm, 21 × 19 × 3.5 cm (length, width and height respectively) were used for keeping the slides of various body parts of crickets.

7) Slides and Cover Slips (Figure 9)

Ordinary slides and cover slips were used for preparing the body parts such as antenna, wings, cerci and genitalia of crickets.

8) Oven (Figure 10)

Oven of size, 3.6 × 2.4 feet (height and width) have been used for drying the slides of crickets and drying killed specimens.

9) Insect Store Box (Figure 11)

Insect store box of size, 47 cm length × 30 cm width was used for keeping insects for longer period.

10) Camera Nikon S4 (Figure 12)

Nikon Coolpix S4 Camera, 10x optical zoom and 6 megapixal was used for taking photographs of crickets.

11) Compound Microscope (Figure 13)

Simple compound microscopes with objectives 10x, 45 x, 100x were used for describing the cricket species.

12) Microphotographs

Microphotographs of whole mounts of the crickets and their various body parts namely antenna, wings, cerci and genitalia have been photographed by using microphotographic camera (Nikon S4).

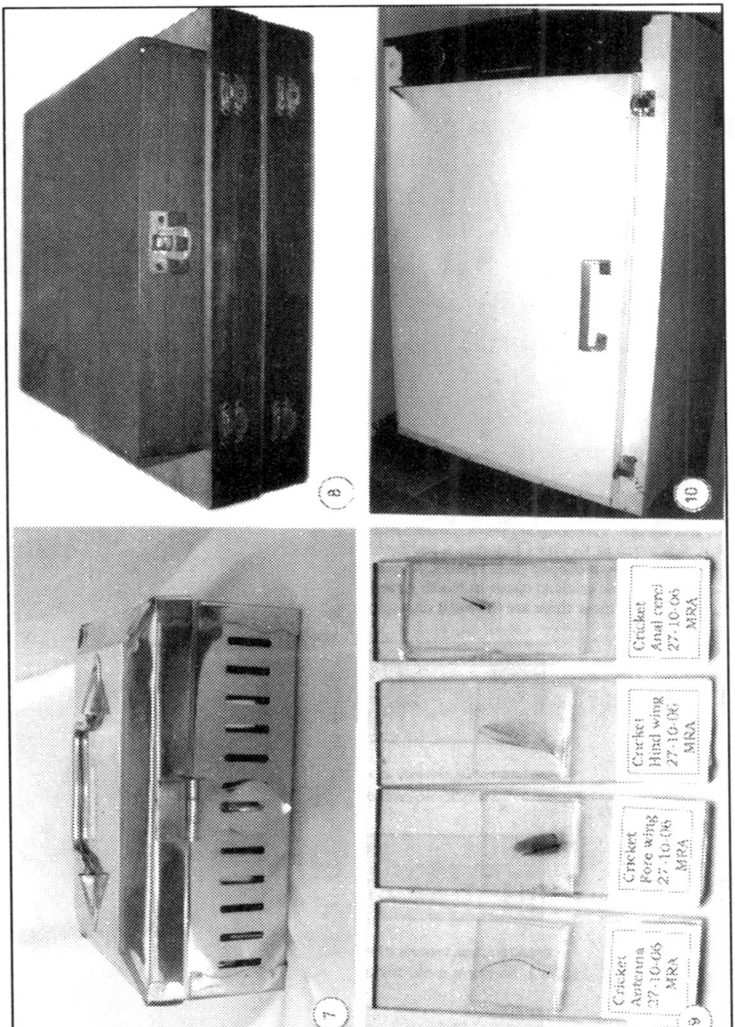

Plate 2

Figure 7: Cricket Rearing Age; Figure 8: Slide Boxes; Figure 9: Slide and Cover Slips; Figure 10: Oven

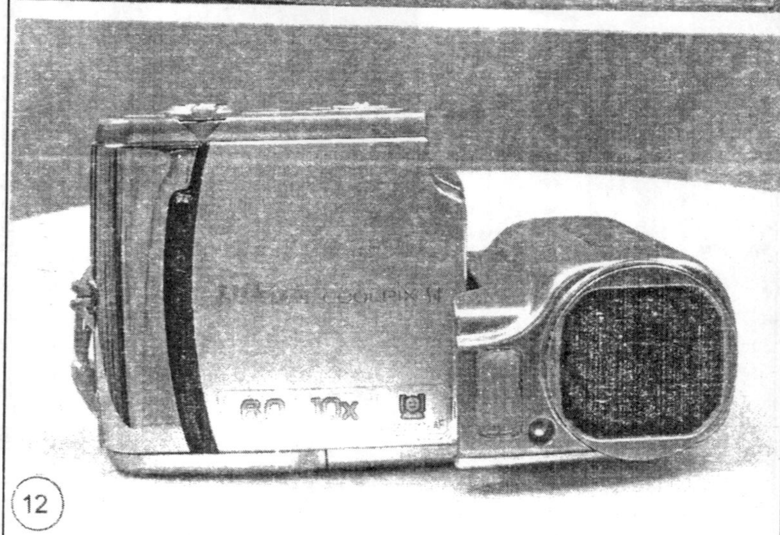

Plate 3
Figure 11: Insect Store Box; Figure 12: Camera Nikon S4

Plate 4
Figure13: Compound Microscope

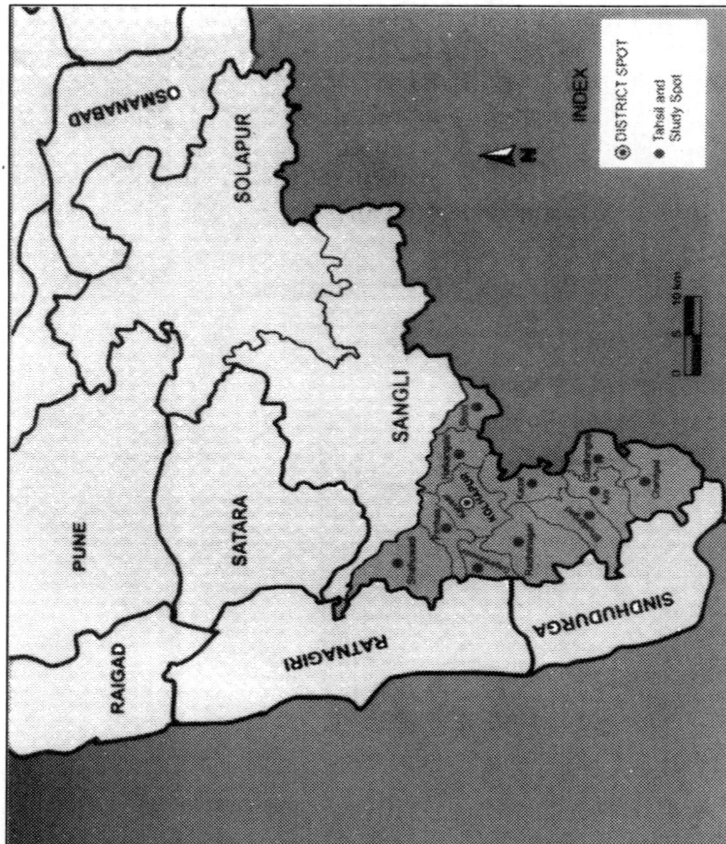

Figure 14: Map of Kolhapur District Showing Study Area

13) Chemicals

Following chemicals were used for preparation of slides and preserving insects.

(*i*) 10 per cent KOH.

(*ii*) 30 per cent to 100 per cent Ethyl alcohol grades.

(*iii*) Glacial acetic acid.

(*iv*) Xylene.

(*v*) DPX/Canada Balsum.

Methods

The cricket species were collected from the district Kolhapur (Figure 14) from the years, 2006 to 2007. Kolhapur lies between 74°.13' East longitude and 16°.41' North latitudes; Kagal lies between 74°.18' and 16°.34' N ; Ajara lies between 74°.28' E and 16°.51' N; Radhanagari lies between 74°.35' and 16°.60' N; Panhala lies between 74°.07' E and 16°.48' N and Shahuwadi lies between 74°.40' E and 16°.58' N. A large number of species were collected by visiting various places of Tahasils of district Kolhapur (Ajra, Radhanagari, Kagal, Kolhapur, Shahuwadi, Panhala, Gaganbawada, Karveer, Hatkangale, Shirol, Bhudargad, Gadhinglaj and Chandgad at 15 days interval. The crickets were collected with the help of insect net (Figure 5) and containers (Figure 6). The nymphs collected from various study spots were reared in the laboratory for their adult formation by using glass troughs (Figure 3) and cricket rearing cages (Figure 7). The adult crickets were narcotized in ether and killed in cyanide killing bottle/chloroform. The crickets were pinned with entomological pins from the dorsal of mesothorax and dried in drying chamber/oven at 55 °C. The dried specimens then kept in wooden insect box. Few naphthol bolls were placed at the bottom of the box for preventing fungal attack. Holotypes and paratypes are pinned, labelled and kept in insect box. The records were made on locality, date of collection and identification. For the taxonomical studies antenna, wings, cerci, genitalia, were mounted on slides in D.P.X. For preparation of slides of genitalia following method (Vasanth, 1993) was adopted. The abdominal extremity was cut open and the genitalia removed. The dissected genitalia was then kept overnight in 10 per cent cold potassium hydroxide solution to clear the muscles. (Keeping for a longer period of time in the KOH solution results in desclerotisation of the genitalic parts, making it extremely difficult for study). Muscles were cleared with the help of needles, the cleared genitalia was then washed thoroughly in distilled water and preserved in 90 per cent alcohol in a microvial. This was either kept along with the specimen (if the specimen were preserved in spirit) or in case of a pinned specimen, it was kept separately and an index-numbered label was given to both the genitalia and specimen from which it was dissected out. Illustrations of the genitalia were made with the help of a camera lucida. The procedure of affixing the genitalia to a piece of paper kept with the pinned specimen was not adopted since this makes it impossible to study the ventral view of the genitalia.

Morphological studies were carried out with the help of monocular microscope. Comparative measurements of body parts of specimens were taken with the help of ocular micrometer and calculated with the help of graduated mechanical stage. All measurements were made in millimeter.

The terminology adopted for description of species is the same as that of Chopard (1928a, 1933a, 1969) and Vasanth (1993) which is represented in the Figures 15 to 24. A large number of references were consulted for the present work and cited in bibliography.

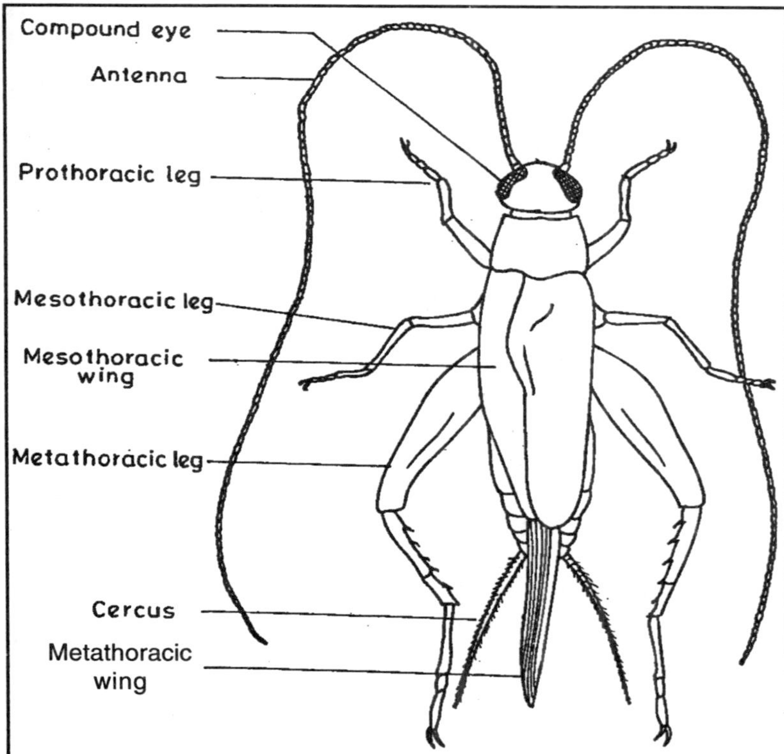

Figure 15: Adult Male (♂)

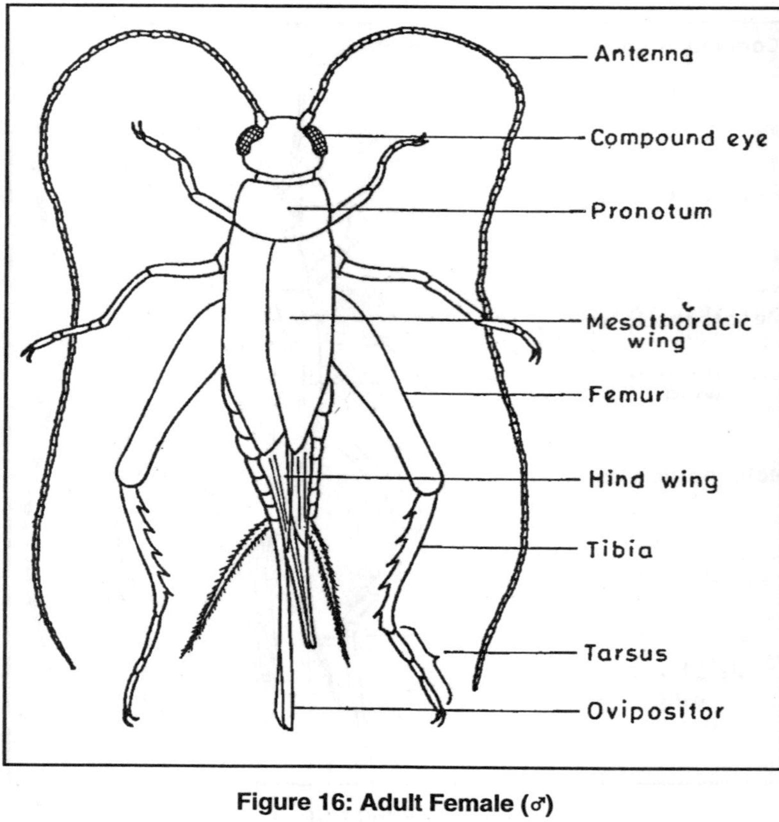

Figure 16: Adult Female (♂)

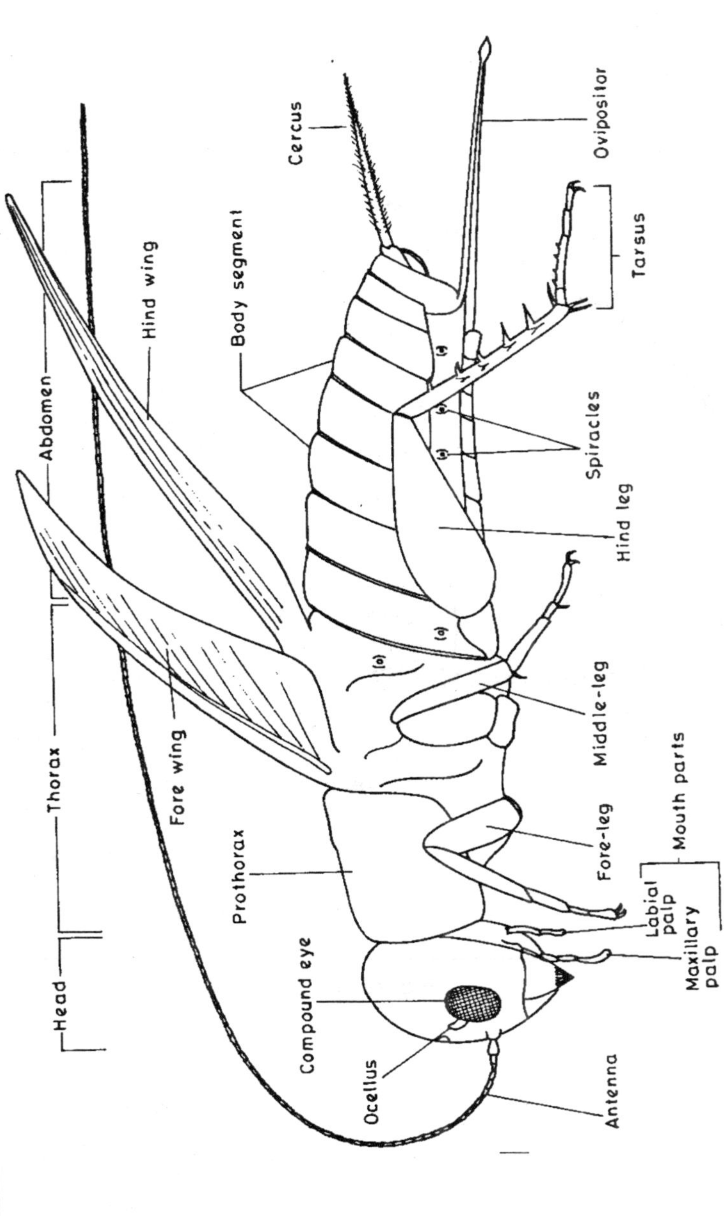

Figure 17: Lateral View of Field Cricket

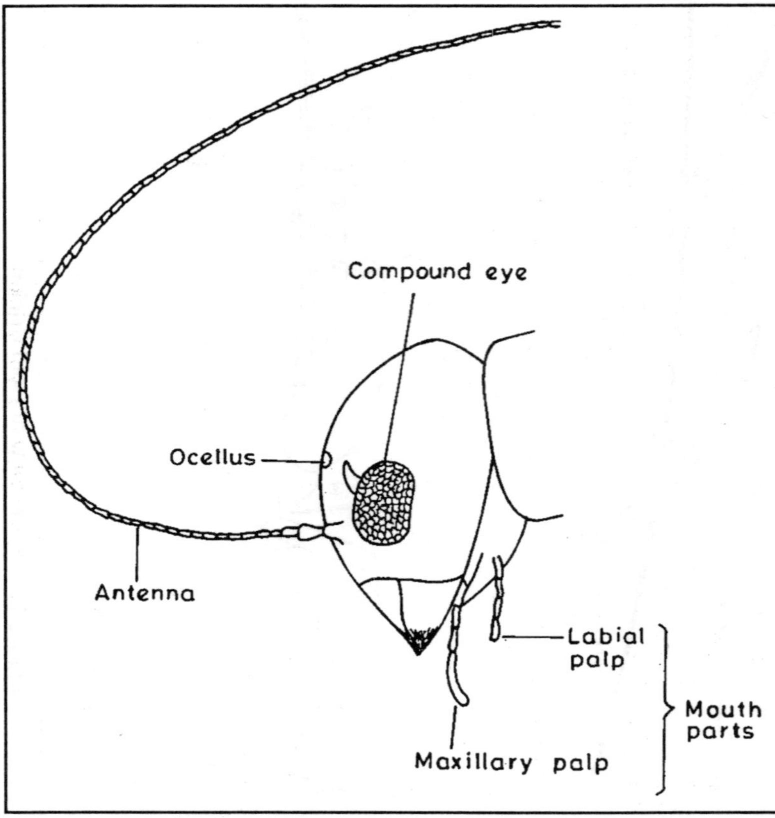
Compound eye
Ocellus
Antenna
Labial palp
Mouth parts
Maxillary palp

Figure 18

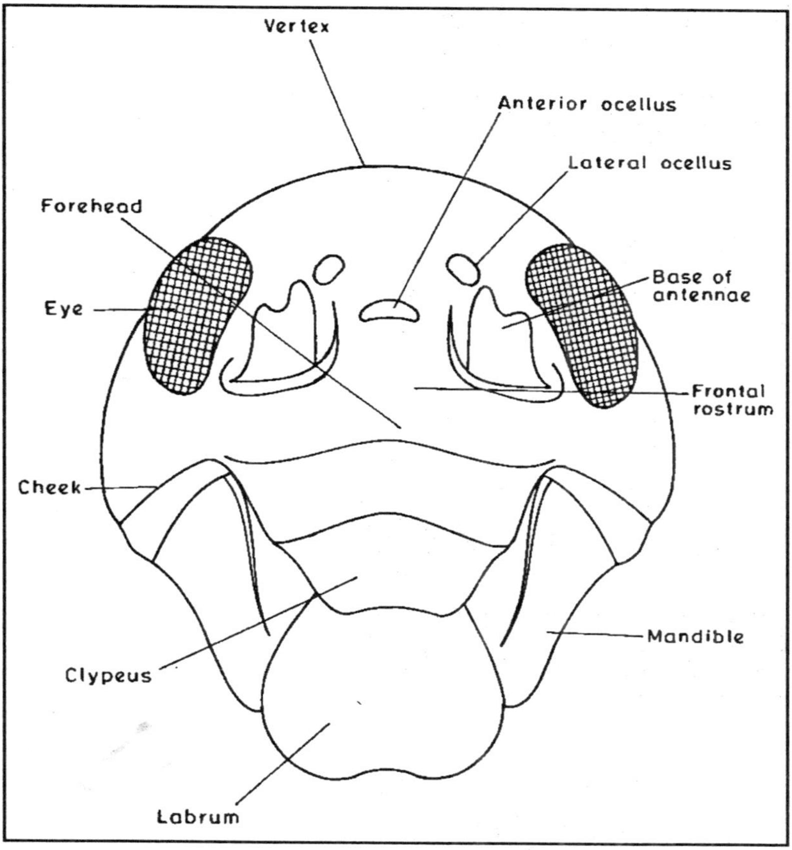

Figure 19: Head of Gryllus

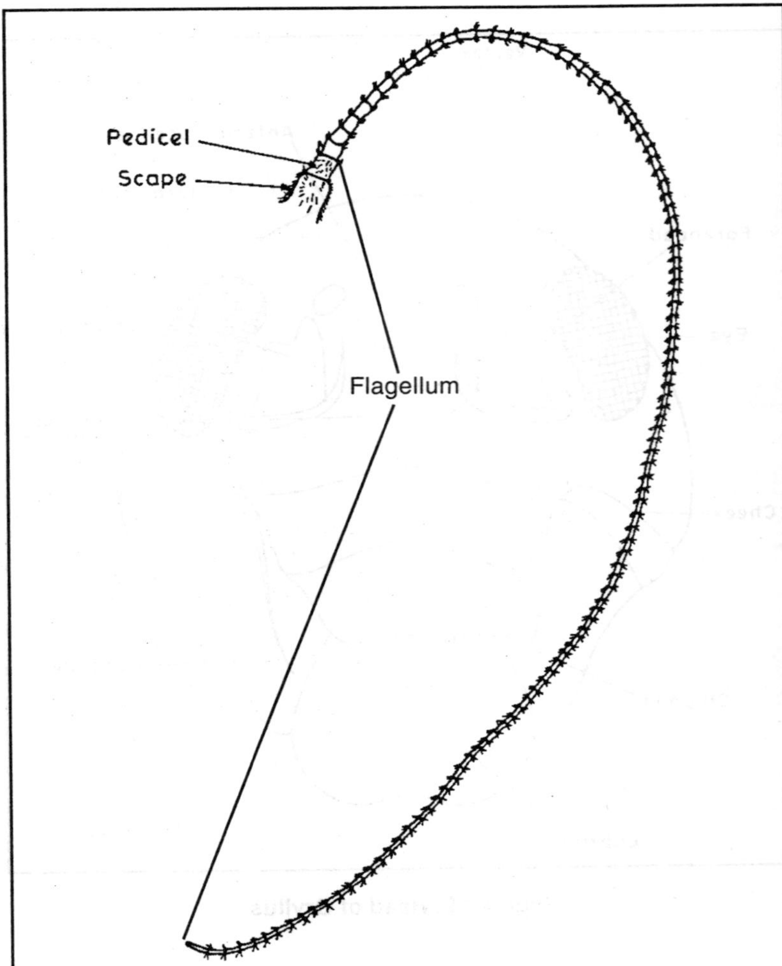

Figure 20: Antenna

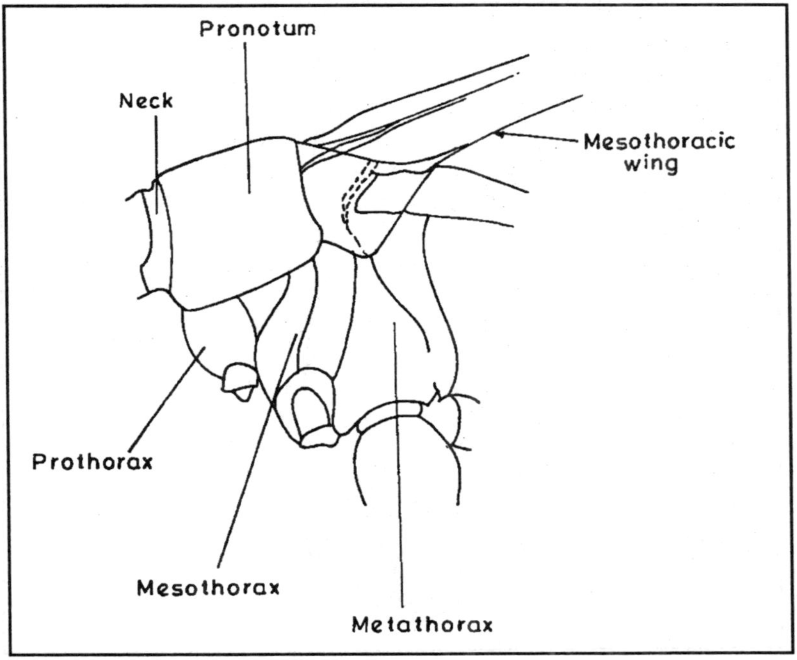

Figure 21: Thorax

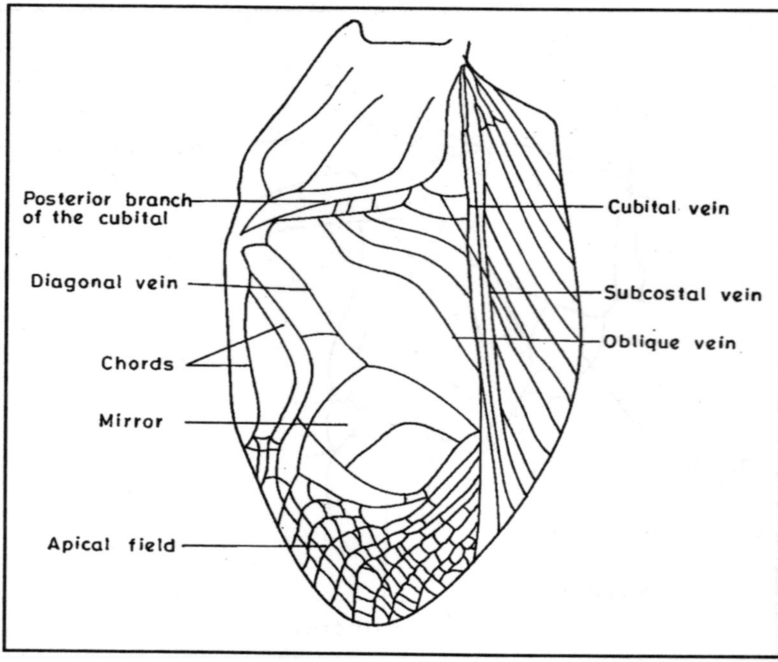

Figure 22: Male Tegmina of *Gryllus*

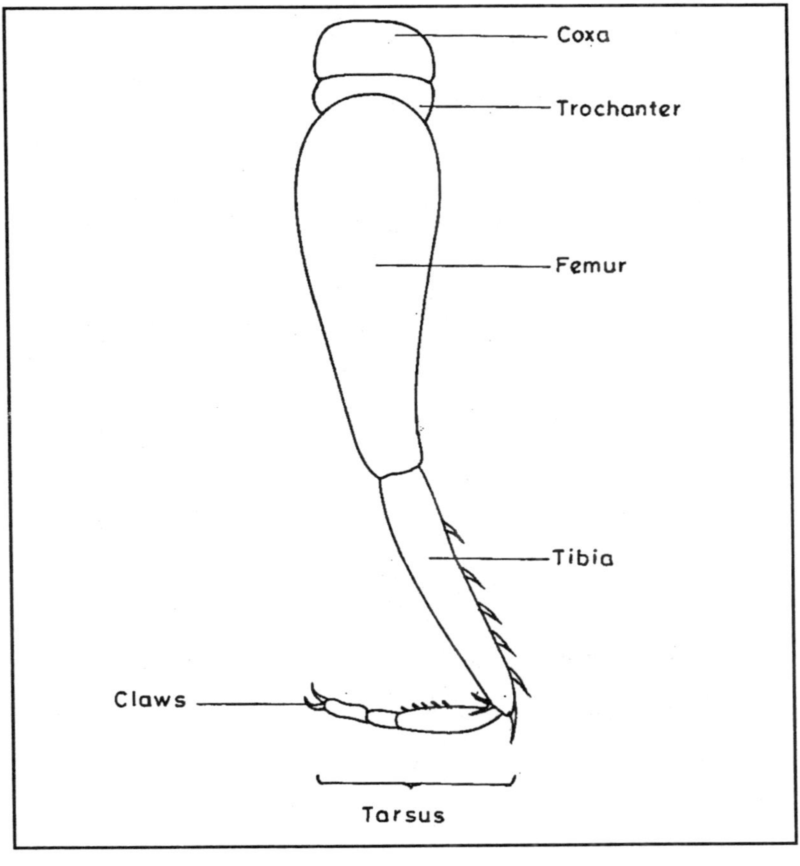

Figure 23: Hind Leg

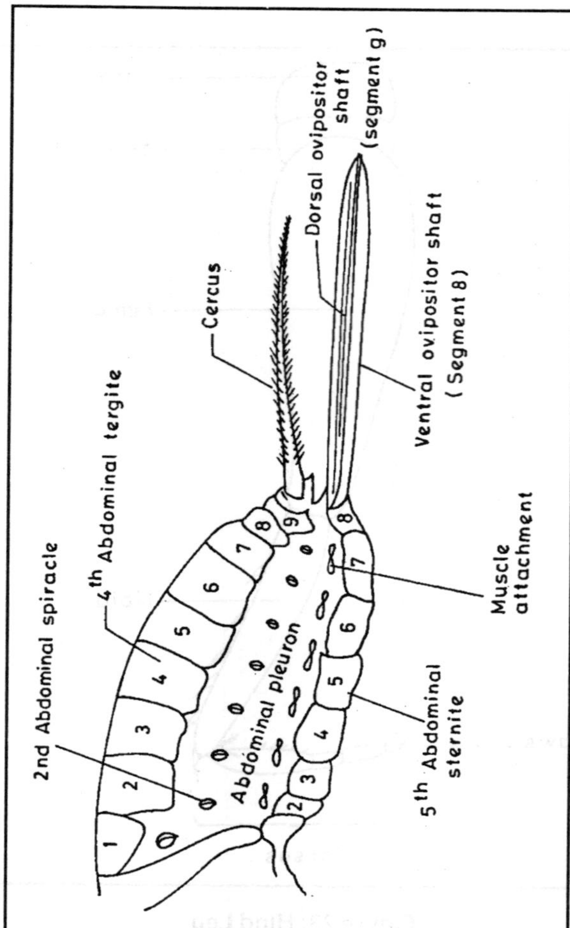

Figure 24: Abdomen

DIVERSITY OF CRICKETS

FAMILY: GRYLLOTALPIDAE

The family Gryllotalpidae is characterized by

1. Elongate body, head small and conical.
2. Ocelli two.
3. Pronotum shape convex.
4. Fore leg with coxa very short, femur somewhat curved; tibia very short, triangular its apical spur transformed into strong, cutting dactyli.
5. There is no mirror and the disk of the tegmina occupied by a large triangular cell.
6. In female ovipositor is completely aborted.

The family Gryllotalpidae is represented by two genera, *Scapteriscus* Scudder and *Gryllotalpa* Latreille.

KEY TO THE GENERA OF GRYLLOTALPIDAE

Anterior tibiae armed with two dactyls .... *Scapteriscus* Scudder

Anterior tibiae armed with four dactyls .... *Gryllotalpa* Latreille

GENUS GRYLLOTALPA LATREILLE

The genus *Gryllotalpa* was erected by Latreille in 1802 with the help of type species *Gryllus* (*Acheta*) *gryllotalpa* L. (*Gryllotalpa* Latreille, *Hist. nat. Crust. Ins.* 111, 275, 1802). The genus is distributed in Europe, Asia, Africa, Australia and Newzeland. From this genus six species have been reported from India (Chopard, 1969; Awate and Sathe, 2007). In past, Chopard, 1928a, 1928b, 1933a, 1933b, 1935, 1951, 1969; Pantel, 1896; Sandrasagara, 1954; Vasanth, 1993; Awate and Sathe, 2007, etc studied the genus from India.

KEY TO THE SPECIES OF GRYLLOTALPA

1. Very large species (40-50mm) .... *hirsuta* Burmeister
 – Species of medium size (20-30 mm) 2
2. Body of a velvet blackish colour with rofous legs .... *fulvipes* Saussure

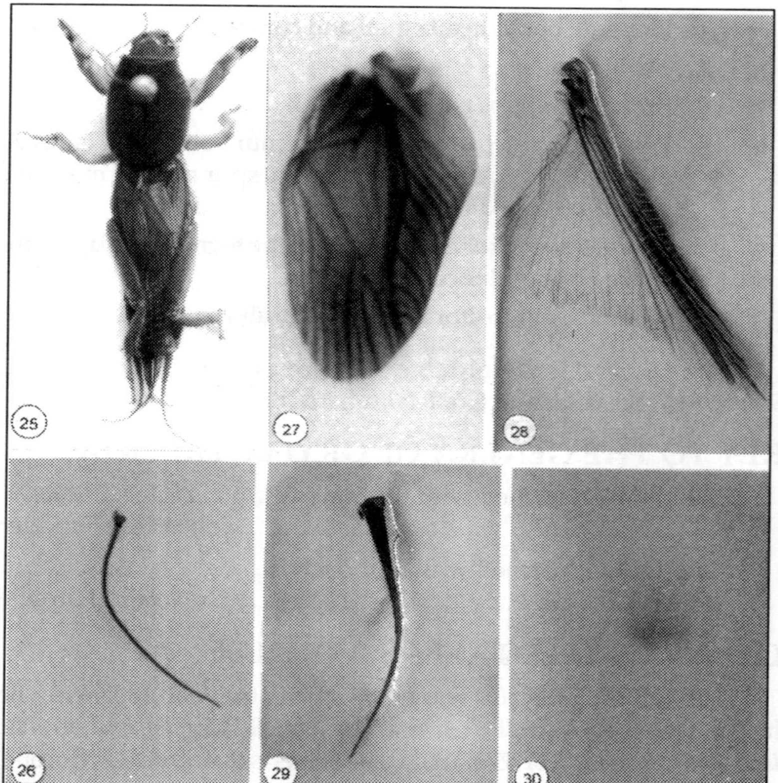

Plate 5: *Gryllotolpa indica* sp.nov.
Figure 25: Adult: Male; Figure 26: Antenna; Figure 27: Tegmina; Figure 28: Hind Wing; Figure 29: Anal Cerci; Figure 30: Genitalia

- Body length more than 30 and less than 40, Ovipositor vestigial .... *vestigi* sp. nov.
- Colour uniformly of a more or less dark brown 3

3. Shape very narrow; dark brown; ♀, veins of tegmina very close, parallel, not at all deviated at base; ♂ mirror very narrow .... *ornata* Walker

- Mirror reduced; tegmina extended to the 3rd segment of abdominal tergite, genitaliastructure, inferior hook short, blunt.. *indica* sp. nov.
- Four spines on mid leg tibia, pronotal plate quadrangular basally, parallel sided, tapering anteriorly .... *shivajiensis* sp. nov.
- Tegmina extended to the 4th abdominal tergite, small sized mirror present .... *micromi* sp. nov.

Gryllotalpa indica sp.nov.

Male (Figure 25)

35.00 mm long excluding anal appendages, body colour uniformly dark brown, head 6.00 mm long, thorax 16.00 mm long, tegmina 14.00 mm long, hind wing 27.00 mm long; abdomen 13.00 mm long, dark brown; anal cerci 11.00 mm long, dark brown.

Head

6.00 mm long, 5.00 mm broad, roundedly triangular; occiput dark brown, hairy; compound eyes black, ocular distance 1.16 mm, interocular distance 2.08 mm; ocelli 2, yellowish, semicircular, ocellar distance 0.34 mm, interocellar distance 0.78 mm; face hairy, antennae on frontal rostrum; clypeus dark brown to blackish, triangular, 1.08 mm long, 1.12 mm broad; labrum dark brown, semicircular, 1.33 mm long, 1.54 mm broad; maxillary palpi dark brown, 4.00 mm long, 0.55 mm broad, 5 segmented.

Antenna (Figure 26)

11.00 mm long, 0.50 mm broad, reddish brown, antennal margin dark brown; scape 0.41 mm long, 0.50 mm broad, hairy; pedicel 0.21 mm long, 0.26 mm broad; flagellum 112 segmented, 10.38 mm long, 0.20 mm broad; 1st segment 0.17 mm long, 0.20 mm broad; 15th segment 0.07 mm long, 0.18 mm broad; terminal segment 0.07 mm long, 0.05 mm broad.

Flagellar Formula

1 = L/W = 0.85, 15 = L/W = 0.38, Ts = L/W = 1.4, A = 0.87

Thoras

16.00 mm long, 8.00 mm broad, dark brown, densely hairy, pronotum with regularly convex side, pronotum not acutely tapering towards anterior, disk and lateral lobes separated by strongly marked humeral edge, lateral lobes rectangular.

Tegmina (Figure 27)

14.00 mm long, 8.00 mm broad, dark brown, short, wide, extended to the 3rd abdominal tergite, mirror very narrow and reduced, oblique vein very much reduced; diagonal vein long, dark brown; subcosta bearing 10 branches.

Hind Wing (Figure 28)

27.00 mm long, 13.00 mm broad, dark brown, wings very long, longer than tegmina.

Legs

Fore leg, hind leg similar to body colour, mid leg lighter than body colour.

Fore Leg

14.60 mm long, 5.00 mm broad; coxa 3.00 mm long, 5.00 mm broad, dark brown, hairy at outer margin; trochanter 2.00 mm long, 2.00 mm broad, dark brown to blackish, curved spine at the tip of trochanter, densely hairy; femur stoughtly builded, 6.00 mm long, 3.00 mm broad, velvet dark brown, densely hairy at outer, margin; tibia 3.00 mm long, 3.00 mm broad, velvet dark brown, densely hairy at outer margin, tibia long, armed, modified into 4 tibial dactyls, dactyls long and narrow, 1st tibial dactyl 0.97 mm long, 0.71 mm broad; 2nd tibial dactyl 1.26 mm long, 0.56 mm broad; 3rd tibial dactyl 3.00 mm long, 0.47 mm broad; 4th tibial dactyl 0.60 mm long, 0.38 mm broad, dark brown, sparsely hairy; claw dark brown.

Mid Leg

20.73 mm long, 4.00 mm broad, coxa 3.00 mm long, 4.00 mm broad, yellowish brown, hairy at outer margin; trochanter 1.73 mm long, 1.65 mm broad, yellowish brown, densely hairy at outer margin; femur stoughtly builded, 6.00 mm long, 1.32 mm broad, yellowish brown, densely hairy; tibia 5.00 mm long, 1.16 mm broad, yellowish brown, narrow, slender, 2 reddish brown tibial spines at tip, outer margin densely hairy; tarsus 3 segmented, 5.00 mm long, 0.48 mm broad, yellowish brown, densely hairy; 1st tarsal segment 1.65 mm long, 0.48 mm broad; 2nd tarsal segment 0.74 mm long, 0.4 mm broad; 3rd tarsal segment 0.74 mm long, 0.34 mm broad; claw reddish dark brown.

Hind Leg

25.04 mm long, 4.00 mm broad; coxa 3.00 mm long, 4.00 mm broad, dark brown, hairy at outer margin; trochanter 1.04 mm long, 1.56 mm broad, yellowish brown, outer margin densely hairy; femur 9.00 mm long, 3.00 mm broad, dark brown, densely hairy; tibia long, narrow, slender, 7.00 mm long, 1.13 mm broad, dark brown, 6 reddish brown tibial spines, densely hairy; tarsus 3 segmented, 5.00 mm long, 0.69 mm broad, dark brown, densely hairy; metatarsus armed with 2 curved spines; 1st tarsal segment 1.39 mm long, 0.55 mm broad; 2nd tarsal segment 0.74 mm long, 0.52 mm broad; 3rd tarsal segment 1.65 mm long, 0.46 mm broad; metatarsus armed with 2 curved spines; claw dark reddish brown.

Abdomen

13.00 mm long, 5.00 mm broad, tapering to posterior side, dark brown, 10 segmented; anal cerci (Figure 29) 11.00 mm long, cerci with long hairs, dark brown.

Genitalia (Figures 30 and 31)

Inferior hooks short, blunt, superior part of pseudepiphallus dark brown, 1.05 mm long, 0.12 nmm broad, inferior part of pseudepiphallus dark brown, 1.29 mm long, 0.27 mm broad, lateral rod dark brown, 0.83 mm long, 0.66 mm broad.

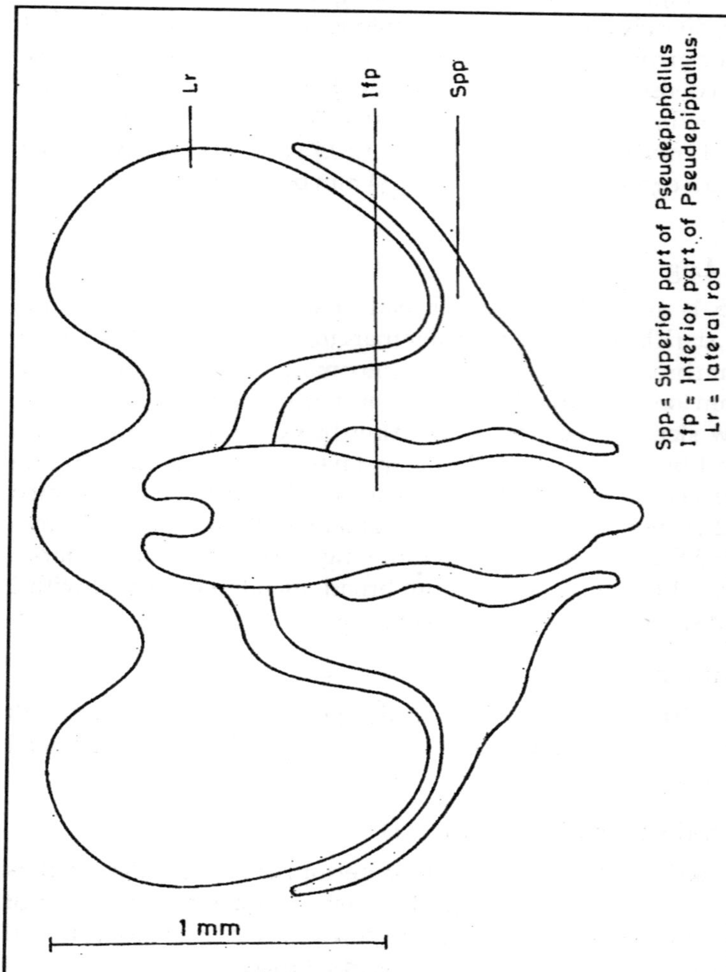

Figure 31: Genitalia of *Gryllotalpa indica* sp.nov.

Colour

Dark Brown

Clypeus, labrum, maxillary palpi, occiput, antennal margin, thorax, tegmina, hind wing, fore leg coxa, claw, hind leg coxa, femur, tibia, genitalia.

Black

Compound eyes.

Dark Brown

Thorax, tegmina, mid leg tibia, hind wing, hind leg coxa, trochanter, femur, abdomen, anal cerci.

Yellowish Brown

Fore leg coxa, mid leg coxa, trochanter, femur, tarsus, hind leg tibia, tarsus.

Reddish Brown

Antennae, spines of fore leg trochanter, mid leg tibial spines, claw.

Velvet Dark Brown

Fore leg femur, tibia.

Dark Reddish to Black

Fore leg tibial dactyls.

Holotype

Male, India, Maharashtra, Coll. 20-X-2006; Awate Manisha; antenna, wings, cerci, genitalia on slide labeled as above; other parts with pinned form, labeled as above.

Paratype

5 ♂, 0 ♀, Coll. from 10-VII-2006 to 10-X-2007.

Distributional Record

1 ♂ Shahuwadi, 6-VII-2006; 2 ♂ Gaganbawada, 24-IX-2006; 1 ♂ Kolhapur, 2-X-2006; 1♂ Bhudargad, 15-VI-2007; 1♂ Karveer, 29-VII-2007.

Etymology

Gryllotalpa indica sp. nov. has been reported from India for the first time, hence the name.

Remarks

According to Chopard's (1969) key *Gryllotalpa indica* sp. nov. runs close to *Gryllotalpa ornata* Walk by having following characters.

1. Body medium sized.
2. Body colour dark brown.
3. Pronotum with regularly convex side.

However, it differs from the above species by having following characters.

1. Tegmina extending to the apex of 3rd segment of abdominal tergite and not 6th abdominal tergite as noticed in *G.ornata.*
2. Subcosta bearing 10 branches.
3. Genitalia structure, inferior hook short, blunt.
4. Tibial dactyls long and narrow.
5. Flagellar formula

   1 = L/W = 0.85, 15 = L/W = 0.38, Ts = L/W = 1.4, A = 0.87
6. Mirror reduced.
7. Cerci with long hairs.
8. Ocelli semicircular.
9. Pronotum not acutely tapering towards anterior.

Gryllotalpa vestigi sp. nov.

Female (Figure 32)

36.00 mm long excluding anal appendages, body velvet blackish, head 6.00 mm long, thorax 15.00 mm long, tegmina 13.00 mm long, hind wing 26.00 mm long; abdomen 15.00 mm long, dark brown; anal cerci 6.00 mm long, dark brown.

Head

6.00 mm long, 4.00 mm broad, roundedly triangular; occiput dark brown; compound eyes dark black, ocular distance 1.00 mm,

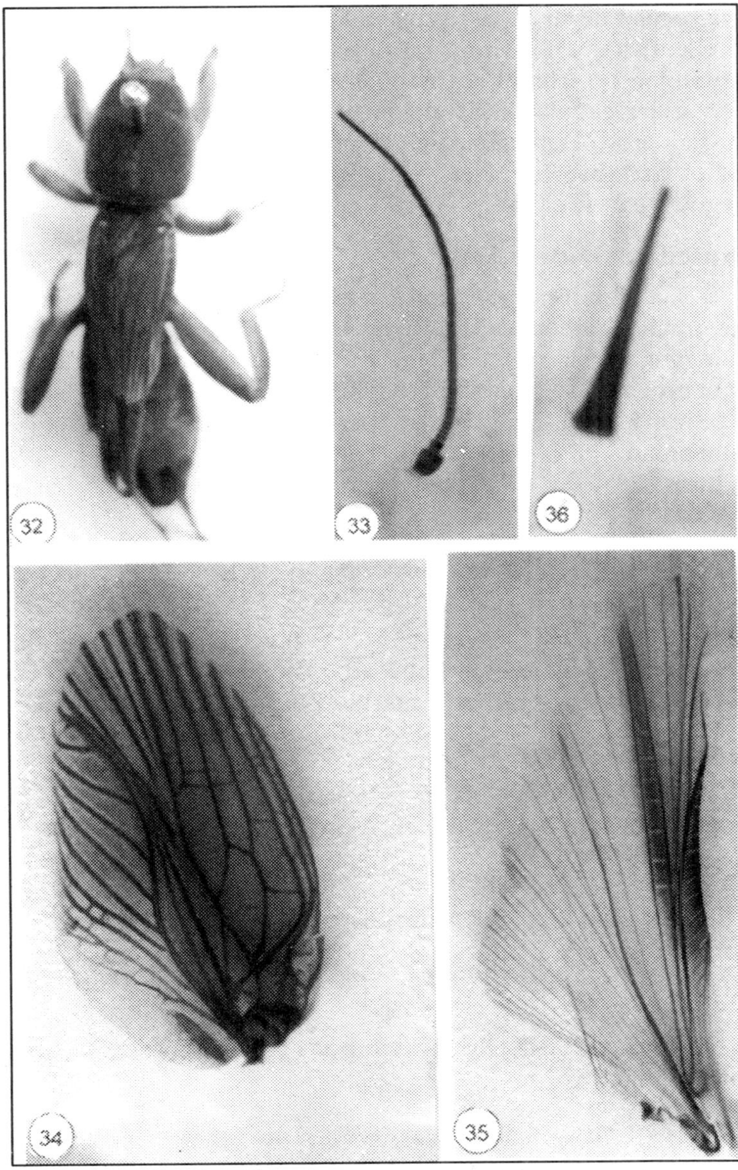

Plate 6: *Gryllotalpa vestigi* sp.nov.
Figure 32: Adult: Female; Figure 33: Antenna; Figure 34: Tegmina; Figure 35: Hind Wing; Figure 36: Anal Cerci

interocular distance 3.00 mm; ocelli 2, yellowish white, semicircular, ocellar distance 0.38 mm, interocellar distance 0.86 mm; face hairy; antennae on frontal rostrum; clypeus dark brown to blackish, triangular, circular, 0.66 mm long, 1.08 mm broad; labrum yellowish to dark brown, semicircular, 0.85 mm long, 1.29 mm broad; maxillary palpi dark brown, cylindrical, 3.00 mm long, 0.33 mm broad, 5 segmented.

Antenna (Figure 33)

9.00 mm long, 0.42 mm broad, dark brown, black margin on antennal segment; scape 0.50 mm long, 0.42 mm broad; pedicel 0.20 mm long, 0.25 mm broad; flagellum 87 segmented, 8.3 mm long, 0.15 mm broad; 1st segment 0.14 mm long, 0.18 mm broad; 15th segment 0.08 mm long, 0.15 mm broad; terminal segment 0.08 mm long, 0.07 mm broad.

Flagellar Formula

1 = L/W = 0.77,15 = L/W = 0.53, Ts = L/W = 1.14, A = 0.81

Thorax

15.00 mm long, 8.00 mm broad, velvet dark brown, pronotum with regularly convex side, broad posteriorly, flat disk and lateral lobes separated by strongly marked humeral edge, rectangular.

Tegmina (Figure 34)

13.00 mm long, 7.00 mm broad, dark brown, long, narrow, extended to the 3rd abdominal tergite; subcosta bearing 8 branches.

Hind Wing (Figure 35)

26.00 mm long, 12.00 mm broad, dark brown, wings very long, longer than tegmina.

Legs

Fore leg, mid leg lighter than body colour; hind leg darker than body colour.

Fore Leg

14.75 mm long, 4.00 mm broad; coxa 3.00 mm long, 4.00 mm broad, yellowish brown; trochanter 3.00 mm long, 2.00 mm broad, reddish brown, hairy, reddish brown spine at tip; femur stoughtly builded, 5.00 mm long, 3.00 mm broad, velvet dark brown, densely hairy at inner margin; tibia 3.00 mm long, 2.00 mm broad, velvet dark brown, densely hairy at inner margin; tibia modified into 4 tibial dactyls, dactyls short and wide; 1st tibial dactyl 0.91 mm long, 0.58 mm broad; 2nd tibial dactyl 1.00 mm long, 0.44 mm broad; 3rd tibial dactyl 1.75 mm long, 0.78 mm broad; 4th tibial dactyl 0.75 mm long, 0.33 mm broad, reddish brown at tip densely hairy; claw reddish brown.

Mid Leg

16.66 mm long, 3.00 mm broad; coxa 2.00 mm long, 3.00 mm broad, yellowish brown; trochanter 2.00 mm long, 2.00 mm broad, yellowish brown, densely hairy; femur 5.00 mm long, 1.05 mm broad, yellowish brown, densely hairy; tibia 5.00 mm long, 2.00 mm broad, dark brown, densely hairy, 4 reddish brown tibial spines; tarsus 3 segmented, 2.66 mm long, 0.33 mm broad, yellowish brown; 1st tarsal segment 1.04 mm long, 0.33 mm broad; 2nd tarsal segment 0.62 mm long, 0.33 mm broad; 3rd tarsal segment 1.00 mm long, 0.20 mm broad; claw dark brown.

Hind Leg

24.00 mm long, 2.00 mm broad, coxa 2.00 mm long, 2.00 mm broad, yellowish brown; trochanter 1.00 mm long, 2.00 mm broad, dark brown, outer margin densely hairy; femur 9.00 mm long, 3.00 mm broad, dark velvet brown, densely hairy; tibia 7.00 mm long, 1.00 mm broad, velvet dark brown, 10 reddish brown tibial spines, densely hairy; tarsus 3 segmented, 5.00 mm long, 0.5 mm broad, yellowish brown; 1st tarsal segment 1.54 mm long, 0.37 mm broad; 2nd tarsal segment 0.69 mm long, 0.45 mm broad; 3rd tarsal segment 1.58 mm long, 0.45 mm broad; claw reddish brown.

Abdomen

15.00 mm long, 7.00 mm broad, tapering to posterior side, dark brown, 9 segmented; anal cerci (Figure 36) 6.00 mm long, hairy, yellowish brown; ovipositor vestigial.

Colour

Dark Brown

Occiput, antenna, maxillary palpi, tegmina, mid leg femur, hind leg trochanter, abdomen.

Dark Black

Compound eyes.

Dark Brown to Blackish

Clypeus

Yellowish White

Ocelli.

Velvet Dark Brown

Pronotum, fore leg femur, tibia, hind leg femur, tibia.

Yellowish Brown

Fore leg coxa, mid leg coxa, trochanter, femur, mid leg tarsus, hind leg coxa, tarsus, anal cerci.

Yellowish to Dark Brown

Labrum.

Reddish Brown

Fore leg trochanter, spine of trochanter, femur, tibia, tibial dactyls, claw, hind leg claw.

Holotype

Female, India, Maharashtra, Coll. 26-2006; Awate Manisha; antenna, wings, cerci on slide labeled as above; other parts with pinned form, labeled as above.

Paratype

10 ♂, 0 ♀, Coll. from 4-VI-2006 to 30-XI-2007.

Distributional Record

2 ♂ Radhanagari, 18-VI-2006; 3 ♂ Panhala, 23-VII-2006; 3 ♂ Ajara, 20–VIII–2007; 1♂ Kagal, 16-IX-2007; 1♂ Karveer, 24-XI-2007.

Etymology

Gryllotalpa vestigi sp. nov. ovipositor in the species is vestigial hence the name.

Remarks

According to Chopard's (1969) key *Gryllotalpa vestigi* sp. nov.

runs close to *Gryllotalpa fulvipes* Sassuure by having following characters.

1. Body velvet blackish.
2. Rufous legs.

However, it differs from the above species by having following characters.

1. Body length longer than *G. fulvipes*.
2. Wing proportion (L : W).
3. Flagellar formula
   1 = L/W = 0.77,15 = L/W = 0.53, Ts = L/W = 1.14, A = 0.81
4. Ovipositor vestigial.
5. Shape and proportion of dactyls on fore leg.

Gryllotalpa shivajiensis sp.nov.

Male (Figure 37)

32.00 mm long excluding anal appendages, body colour not uniformly dark brown, head 5.00 mm long, thorax 14.00 mm long, tegmina 12.00 mm long, hind wing 27.00 mm long; abdomen 13.00 mm long, dark brown; anal cerci 11.00 mm long, yellowish brown to dark brown.

Head

5.00 mm long, 5.00 mm broad, roundedly triangular; occiput dark brown to black, hairy; compound eyes dark black, ocular distance 1.00 mm, interocular distance 2.00 mm; ocelli 2, yellowish, semicircular, ocellar distance 0.38 mm, interocellar distance 0.71 mm; face hairy; antennae on frontal rostrum, clypeus dark brown, triangular, 1.00 mm long, 0.58 mm broad; labrum, yellowish brown, circular, 1.75 mm long, 1.29 mm broad; maxillary palpi yellowish brown, cylindrical, 4.00 mm long, 0.30 mm broad, 5 segmented.

Antenna (Figure 38)

12.00 mm long, 0.44 mm broad, yellowish to dark brown, antennal margin dark brown; scape 0.51 mm long, 0.44 mm broad, hairy; pedicel 0.19 mm long, 0.25 mm broad; flagellum 110 segmented, 11.3 mm long, 0.16 mm broad; 1st segment 0.12 mm

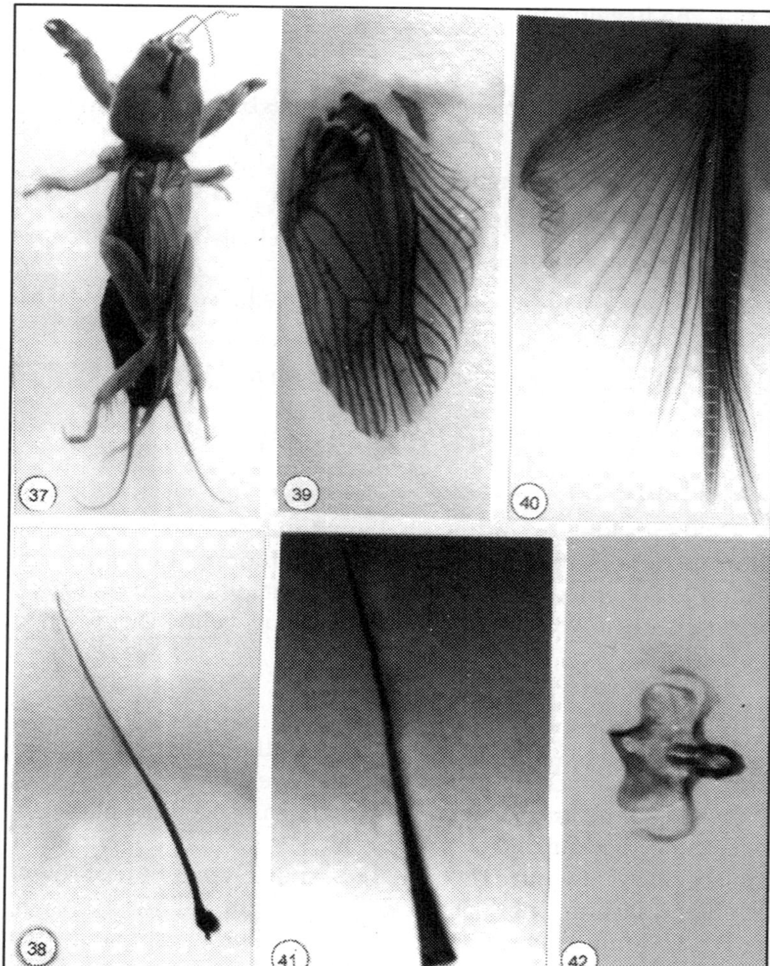

Plate 7: *Gryllotalpa shivajiensis* sp.nov.
Figure 37: Adult: Male; Figure 38: Antenna; Figure 39: Tegmina; Figure 40: Hind Wing; Figure 41: Anal Cerci; Figure 42: Genitalia

long, 0.16 mm broad; 15th segment 0.08 mm long, 0.15 mm broad; terminal segment 0.06 mm long, 0.05 mm broad.

Flagellar Formula

1 = L/W = 0.75,15 = L/W = 0.53, Ts = L/W = 1.2, A = 0.82

Thorax

14.00 mm long, 7.00 mm broad, dark brown, densely hairy, pronotum feebly convex, pronotal plate more or less quadrangular, parallel sided, basally tapering anteriorly; disk and lateral lobes separated by strongly marked humeral edge, lateral lobes rectangular.

Tegmina (Figure 39)

12.00 mm long, 7.00 mm broad, long, narrow, extended to the 3rd abdominal tergite, mirror very much reduced, dark brown; subcosta bearing 9 branches.

Hind Wing (Figure 40)

27.00 mm long, 12.00 mm broad, dark brown, wings very long, longer than tegmina.

Legs

Fore leg, mid leg lighter than body colour; hind leg darker than body colour.

Fore Leg

15.37 mm long, 3.00 mm broad; coxa 3.00 mm long, 3.00 mm broad, yellowish brown, hairy at outer margin; trochanter 2.00 mm long, 2.00 mm broad, dark brown to blackish, dark brown spine at the tip of trochanter, densely hairy; femur stoughtly builded, 6.00 mm long, 3.00 mm broad, velvet dark brown, densely hairy at outer, margin; tibia 4.00 mm long, 3.00 mm broad, velvet light dark brown, densely hairy at outer margin; tibia long, armed, modified into 4 tibial dactyls; dactyls long and narrow, dark reddish to black at tip; 1st tibial dactyl 0.90 mm long, 0.54 mm broad; 2nd tibial dactyl 1.4 mm long, 0.66 mm broad; 3rd tibial dactyl 2.1 mm long, 0.70 mm broad; 4th tibial dactyl 0.74 mm long, 0.4 mm broad, dark brown, sparsely hairy; claw reddish brown.

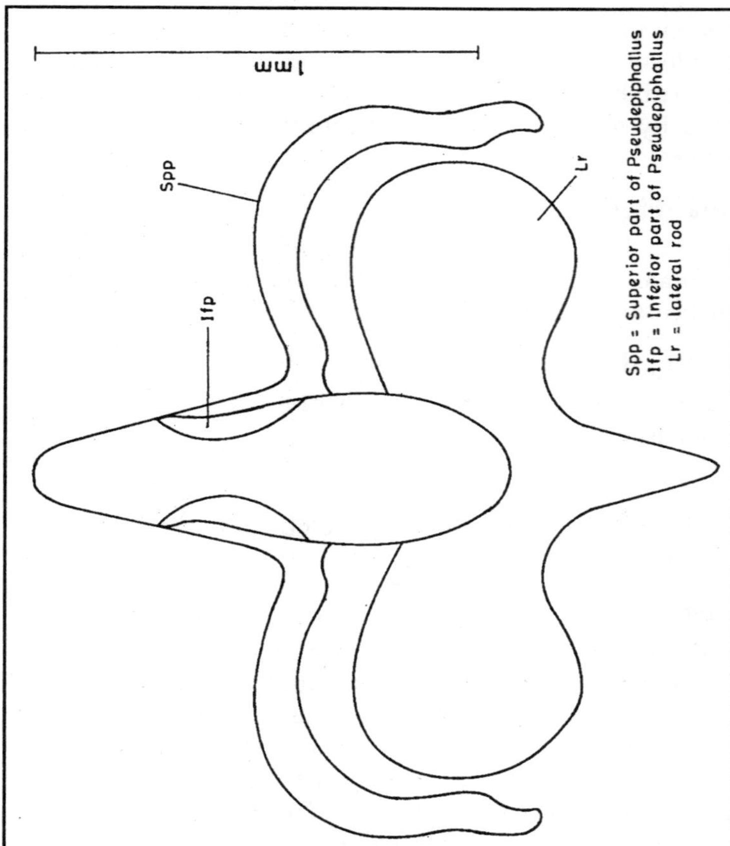

Figure 43: Genitalia of *Gryllotalpa shivajiensis* sp.nov.

Mid Leg

17.00 mm long, 2.00 mm broad; coxa 2.00 mm long, 2.00 mm broad, yellowish brown, hairy at outer margin; trochanter 2.00 mm long, 1.00 mm broad, yellowish brown, sparsely hairy; femur stoughtly builded, 5.00 mm long, 1.12 mm broad, dark brown, densely hairy; tibia narrow, slender 4.00 mm long, 1.00 mm broad, 4 dark reddish brown tibial spines at tip, outer margin densely hairy; tarsus 3 segmented, 4.00 mm long, 0.32 mm broad, yellowish brown, hairy; 1st tarsal segment 2.00 mm long, 0.44 mm broad; 2nd tarsal segment 1.2 mm long, 0.35 mm broad; 3rd tarsal segment 0.5 mm long, 0.47 mm broad; claw reddish brown.

Hind Leg

23.00 mm long, 3.00 mm broad; coxa 2.00 mm long, 3.00 mm broad, dark brown, outer margin densely hairy; trochanter 2.00 mm long, 2.00 mm broad, dark brown, outer margin densely hairy; femur 9.00 mm long, 3.00 mm broad, dark brown to blackish, densely hairy; tibia long, narrow, slender, 5.00 mm long, 0.95 mm broad, yellowish brown, 10 reddish brown tibial spines, densely hairy; tarsus 3 segmented, 5.00 mm long, 0.41 mm broad, yellowish brown, densely hairy; 1st tarsal segment 2.00 mm long, 0.5 mm broad; 2nd tarsal segment 1.00 mm long, 0.46 mm broad; 3rd tarsal segment 2.00 mm long, 0.41 mm broad; claw dark reddish brown.

Abdomen

13.00 mm long, 6.00 mm broad, tapering to posterior side, dark brown to blackish, 10 segmented; anal cerci (Figure 41) 11.00 mm long, hairy, yellowish brown to dark brown.

Genitalia (Figures 42 and 43)

Superior part of pseudepiphallus, yellowish brown, 1.33 mm long, 0.16 mm broad; inferior part of pseudepiphallus dark brown, 1.16 mm long, 0.20 mm broad; lateral rod yellowish brown, 1.62 mm long, 0.58 mm broad.

Colour

Dark Brown to Black

Occiput, fore leg trochanter, abdomen.

Black

Compound eyes.

Yellowish to Dark Brown

Antennae, anal cerci.

Dark Brown

Clypeus, thorax, tegmina, mid leg tibia, hind wing, spine of fore leg trochanter, hind leg coxa, trochanter, femur, 4th tibial dactyl.

Yellowish Brown

Labrum, maxillary palpi, fore leg coxa, mid leg coxa, trochanter, femur, tarsus, hind leg tibia, tarsus, genitalia.

Reddish Brown

Mid leg tibial spines, claw.

Velvet Dark Brown

Fore leg femur, tibia.

Dark Reddish to Black

1st, 2nd, 3rd fore leg tibial dactyl.

Holotype

Male, India, Maharashtra, Coll. 19-X-2006; Awate Manisha; antenna, wings, cerci, genitalia on slide labeled as above; other parts with pinned form, labeled as above.

Paratype

5 ♂, 0 ♀, Coll. from 10-VII-2006 to 10–X–2007.

Distributional Record

1♂ Gaganbawda, 15-VIII-2006; 1♂ Hatkangale, 31-VIII-2006; 1♂ Kolhapur, 5-IX-2006; 2♂ Radhanagari, 3-VI–2007.

Etymology

The name *Gryllotalpa shivajiensis* sp. nov. given to the species is the honour to our beloved institute, Shivaji University, Kolhapur.

Remarks

According to the key of Chopard's (1969) *Gryllotalpa shivajiensis* sp. nov. runs close to *Gryllotalpa ornata* Walk., the species also runs close to *Gryllotalpa indica* sp. nov. by having following characters.

1. Body medium sized.
2. Body colour not uniformly dark brown.
3. Tegminal venation.
4. Clypeus and thorax dark brown.

However, it differs from above two species by having following characters.

1. Tibial proportation length × width.
2. 1st tibial dactyl proportation length × width.
3. 3rd tibial dactyl proportation length × width.
4. Mid leg trochanter proportation length × width.
5. Mid leg tibial spines number (four spines present instead of two).
6. 1st tarsal segment proportation length × width.
7. Abdominal proportation length × width.
8. Genitalia lateral rod proportation length × width.
9. Pronotal plate more or less quadrangular, parallel sided, basally tapering anteriorly.
10. Flagellar formula

    1 = L/W = 0.75,15 = L/W = 0.53, Ts = L/W = 1.2, A = 0.82
11. Subcosta bearing 9 branches.

Gryllotalpa micromi sp.nov.

Male (Figure 44)

31.00 mm long excluding anal appendages, body more or less uniformly dark brown, shape more narrow, head 4.00 mm long, thorax 15.00 mm long, tegmina 13.00 mm long, hind wing 26.00 mm long; abdomen 12.00 mm long, dark brown; anal cerci 9.00 mm long, yellowish brown.

Head

4.00 mm long, 4.00 mm broad, roundedly triangular, occiput dark brown, hairy, compound eyes dark black, ocular distance 0.75

mm, interocular distance 2.00 mm; ocelli 2, yellowish, semicircular, ocular distance 0.26 mm, interocellar distance 0.60 mm; antennae on frontal rostrum, face hairy; clypeus blackish, semicircular, 0.66 mm long, 0.70 mm broad; labrum dark brown, semicircular, 0.68 mm long, 1.01 mm broad; maxillary palpi dark brown, cylindrical, 4.00 mm long, 0.50 mm broad, 5 segmented.

Antenna (Figure 45)

10.00 mm long, 0.38 mm broad, yellowish to dark brown, antennal margin dark brown; scape 0.36 mm long, 0.45 mm broad, hairy; pedicel 0.22 mm long, 0.29 mm broad; flagellum 97 segmented, 9.42 mm long, 0.18 mm broad; 1st segment 0.14 mm long, 0.18 mm broad; 15th segment 0.07 mm long, 0.14 mm broad; terminal segment 0.07 mm long, 0.05 mm broad.

Flagellar Formula

$1 = L/W = 0.77, 15 = L/W = 0.50, Ts = L/W = 1.40, A = 0.89$

Thorax

15.00 mm long, 7.00 mm broad, velvet dark brown, densely hairy, pronotum with regularly convex side; lateral lobes separated by strongly marked humeral edge, triangular.

Tegmina (Figure 46)

Small, narrow, extended to the 4th abdominal tergite, small sized mirror, 13.00 mm long, 7.00 mm broad, dark brown, subcosta bearing 9 branches.

Hind Wing (Figure 47)

26.00 mm long, 11.00 mm broad, dark brown, wings very long, longer than tegmina.

Legs

Fore legs, mid legs lighter than body colour, hind legs darker than body colour.

Fore Leg

13.00 mm long, 3.00 mm broad; coxa 3.00 mm long, 3.00 mm broad, hairy, yellowish brown; trochanter 2.00 mm long, 2.00 mm broad, yellowish brown, reddish brown spine at the tip of trochanter,

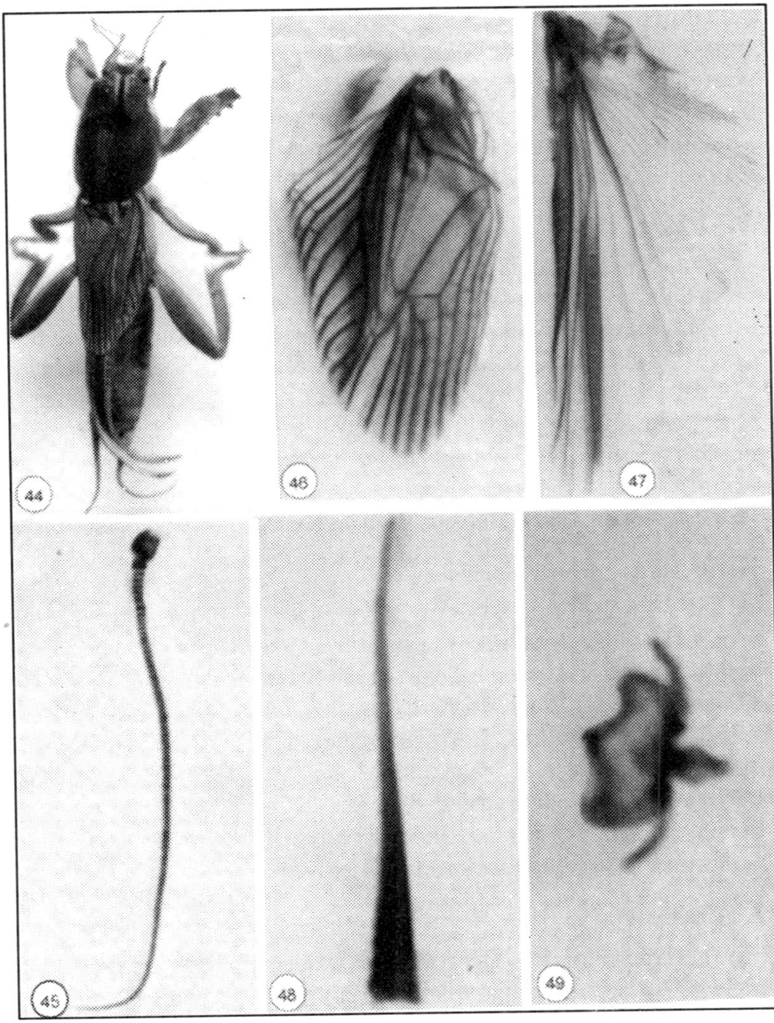

Plate 8: *Gryllotalpa micromi* sp.nov.
Figure 44: Adult: Male; Figure 45: Antenna; Figure 46: Tegmina; Figure 47: Hind Wing; Figure 48: Anal Cerci; Figure 49: Genitalia

densely hairy; femur stoughtly builded, 5.00 mm long, 3.00 mm broad, velvet dark brown, densely hairy at outer margin; tibia 3.00 mm long, 3.00 mm broad, velvet dark brown, densely hairy, tibia long, armed, modified into 4 tibial dactyls, dactyls long and narrow; 1st tibial dactyl 0.89 mm long, 0.58 mm broad; 2nd tibial dactyl 1.2 mm long, 0.60 mm broad; 3rd tibial dactyl 2.10 mm long, 0.75 mm broad; 4th tibial dactyl 0.68 mm long, 0.40 mm broad, bifurcated, dark reddish brown; claw reddish brown.

Mid Leg

16.47 mm long, 3.00 mm broad; coxa 2.00 mm long, 3.00 mm broad, yellowish brown, hairy; trochanter 1.47 mm long, 1.56 mm broad, yellowish brown, hairy; femur stoughtly builded, 4.00 mm long, 1.13 mm broad, dark brown, densely hairy; tibia 5.00 mm long, 1.04 mm broad, narrow, slender, 3 dark reddish brown tibial spines at tip, dark brown, hairy; tarsus 3 segmented, 4.00 mm long, 0.46 mm broad, yellowish brown, hairy; 1st tarsal segment 2.00 mm long, 0.5 mm broad; 2nd tarsal segment 0.37 mm long, 0.45 mm broad; 3rd tarsal segment 1.33 mm long, 0.39 mm broad; claw reddish brown.

Hind Leg

19.69 mm long, 2.00 mm broad; coxa 2.00 mm long, 2.00 mm broad, dark brown, hairy; trochanter 0.69 mm long, 1.04 mm broad, yellowish brown, outer margin densely hairy; femur 7.00 mm long, 3.00 mm broad, dark brown, densely hairy; tibia long, narrow, slender, 6.00 mm long, 1.00 mm broad, dark brown, 8 tibial spines, densely hairy; tarsus 3 segmented, 4.00 mm long, 0.60 mm broad, dark brown, densely hairy; 1st segment 1.77 mm long, 0.60 mm broad; 2nd segment 0.69 mm long, 0.45 mm broad; 3rd segment 1.2 mm long, 0.48 mm broad; claw dark reddish brown.

Abdomen

12.00 mm long, 6.00 mm broad, tapering to posterior side, dark brown, 10 segmented; anal cerci (Figure 48) 10.00 mm long, hairy, yellowish brown.

Genitalia (Figure 49 and 50)

Superior part of pesudepiphallus 1.56 mm long, 0.08 mm broad, yellowish brown; inferior part of pesudepiphallus 0.5 mm long, 0.3

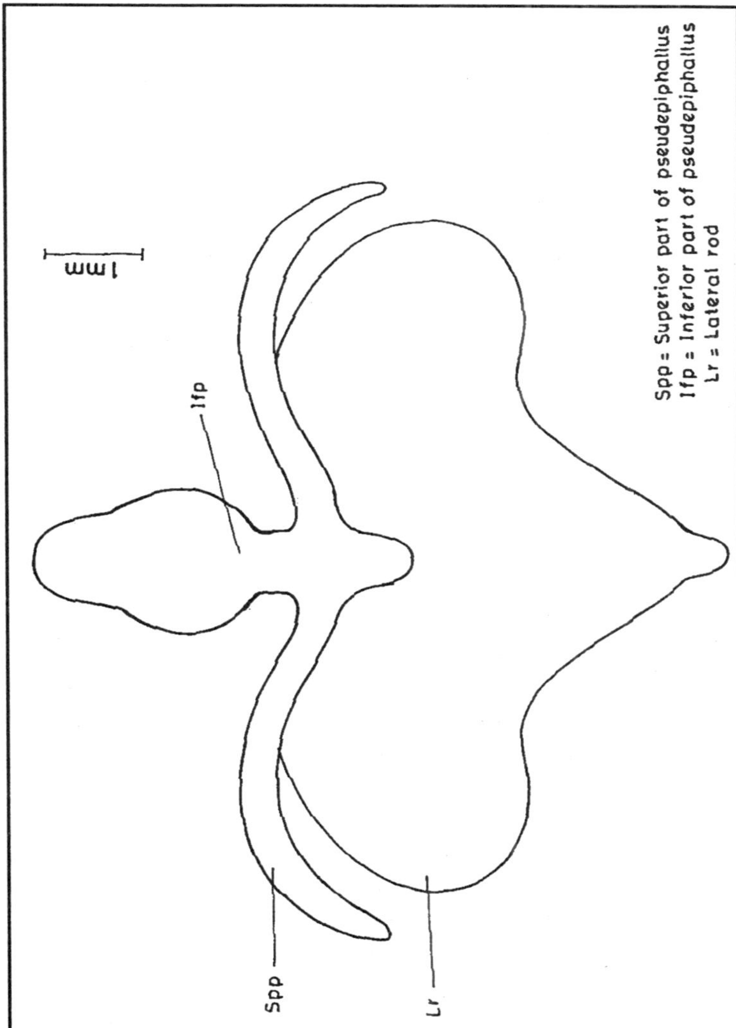

Figure 50: Genitalia of *Gryllotalpa micromi* sp.nov.

mm broad, yellowish brown; lateral rod 1.00 mm long, 0.41 mm broad.

Colour

Dark Brown

Occiput, labrum, maxillary palpi, tegmina, hind wing,mid leg femur, tibia, hind leg coxa, femur, tibia, tarsus, abdomen.

Black

Compound eyes.

Blackish

Clypeus

Yellowish

Ocelli.

Yellowish to Dark Brown

Antennae.

Velvet Dark Brown

Thorax, fore leg femur, tibia.

Yellowish Brown

Fore leg coxa, trochanter, mid leg coxa, trochanter, tarsus, hind leg trochanter, anal cerci, genitalia.

Reddish Brown

Claw.

Holotype

Male, India, Maharashtra, Coll.10-X-2006; Awate Manisha; antenna, wings, cerci, genitalia on slide labelled as above; other parts with pinned form, labelled as above.

Paratype

6 ♂, 0 ♀, Coll. from 3-VI-2006 to 10-X–2007.

Distributional Record

1♂ Kagal, 29-VI–2006; 1♂ Gaganbawada, 15-VIII-2006; 1♂ Kolhapur, 10-X-2006; 2♂ Ajara, 26-XI-2006; 1♂ Shirol, 8-VII-2007.

Etymology

The name *Gryllotalpa micromi* sp. nov. given to the species is due to the presence of micromirror.

Remarks

According to the key of Chopard's (1969) *Gryllotalpa micromi* sp. nov. runs close to *Gryllotalpa ornata* Walk. by having following characters.

1. Body colour more or less uniformly dark brown.
2. Occiput dark brown, compound eyes black.

However, it differs from above species by having following characters.

1. Pronotal plate regularly convex side.
2. Tegmina extended to the 4th abdominal tergite.
3. Small sized mirror present.
4. Dactyl structures of fore leg tibia.
5. Trochanter proportion length × width.
6. Three dark reddish brown tibial spines.
7. Genitalia proportion length × width.
8. Flagellar formula

   1 = L/W = 0.77,15 = L/W = 0.50, Ts = L/W = 1.40, A = 0.89.
9. Subcosta bearing 9 branches.

FAMILY: GRYLLIDAE

The family Gryllidae is characterized by following features.

1. Head big and globular.
2. Tarsi compressed.
3. Hind tibiae armed with spines but never serrulated between these spines.

The family Gryllidae is subdivided into two subfamilies *viz.*, Gryllinae and Nemobiinae. The subfamily Gryllinae is characterized by spines immovable on hind tibia and without hairs. The subfamily

Gryllinae contains 23 genera namely, *Landreva, Ectolandrevus, Landrevus, Gryllus, Brachytrupes, Acheta, Liogryllus, Gymnogryllus, Scapsipedus, Velarifictorus, Loxoblemmus, Platyblemmus, Stephoblemmus, Parasciobia, Gryllodes, Gryllopsis, Turanogryllus, Cophogryllus, Modicogryllus, Lenigryllus, Plebeiogryllus, Platygryllus, Teleogryllus.*

SUBFAMILY GRYLLINAE

Diagnostic Characters

1. Blackish, brownish or tawny coloured crickets of variable size.
2. Indiviuals with globular head.
3. Minute and compressed second tarsal segment with strong, glabrous, immovable spines.
4. Serrulation in between spines on posteriortibiae absent.
5. Both pairs of wings generally moderately well to fully developed or reduced.

KEY TO THE INDIAN GENERA OF SUBFAMILY GRYLLINAE (Vasanth, 1993)

1. Posterior tibiae presenting serrulation proximal to spines 2
– Posterior tibiae without serrulation 4
2. Anterior tibiae without tympanum .... *Chopardia* Bhowmik
– Anterior tibiae with at least one tympanum 3
3. Anterior tibiae presenting a tympanum on each face .... *Doulandrevus* Kirby
– Anterior tibiae presenting a tympanum on external face only .... *Landreva* Walker
4. Body glabrous or nearly so; ocelli . arranged in a straight line; length of posterior femora at least equal to the combined lengths of tibia and tarsus 5
– Body usually more pubescent; ocelliarranged in the form of a triangle; length ofposterior femora never greater than combined lengths of tibia and tarsus 7

5. Pronotum greatly widening in front, with anterior margin strongly concave; abdomen narrow at base; female tegmina very short .... *Sphecogryllus* Chopard
– Pronotum only gradually widening in front with anterior margin feebly concave; abdomen not narrow at base; female tegmina fully developed 6
6. Large, robust insects; male pronotum widening more strongly in front, anterior metatarsi very short, anterior and middle tibiae not ciliated with long hairs; ovipositor very short in relation to body size .... *Brachytrupes* Serville
– Smaller and slender insects; male pronotum, if at all, very slightly widening in front; anterior metatarsi moderately long, anterior and middle tibiae ciliated with long hairs; ovipositor longer in relation to body size .... *Gymnogryllus* Saussuer
7. Supra–anal plate bears a prominent pair of posteriorly projecting, heavily sclerotised structure .... *Spinogryllys* Vasanth
– Supra-anal plate bears no such structures 8
8. Face of male flattened, or more or less concave; frontal rostrum with or without prolongation or process 9
– Face of male convex; frontal rostrum never with prolongation or process 15
9. Both sexes apterous .... *Scapsipedoides* Chopard
– Tegmina present in both sexes in various degrees of development 10
10. Face of male flattened, or slightly concave 11
– Face of male strongly flattened; frontal rostrum prolonged, and may be provided with a process 12
11. Frontal rostrum of male subangulate, prominent, with a small apical depression .... *Coiblemmus* Chopard
– Frontal rostrum of male convex, with no special angulation or apical depression .... *Velarifictorus* Randell
12. Frontal rostrum of male projecting, simply angular, with no process .... *Loxoblemmus* Saussure
– Frontal rostrum of male strongly projecting a corneous or membranaceous process 13

13. Frontal rostrum of male with a wide, corneous process which is never membranaceous.....*Stephoblemmus* Saussure
 – Frontal rostrum of male with an entire or divided membranaceous process 14
14. Membranaceous process of frontal rostrum deeply divided into two large lobes .... *Meristoblemmus* Chopard
 – Membranaceous process of frontal rostrum entire .... *Parasciobia* Chopard
15. Width of frontal rostrum subequal to that of first antennal segment 16
 – Width of frontal rostrum always about double that of first antennal segment 17
16. Head dorsally flattened; ovipositor long .... *Gryllodes* Saussure
 – Head dorsally rounded; ovipositor rudimentary .... *Itaropsis* Chopard
17. Body convex, cylindrical; female with imperfectly developed tegmina, often reduced to lateral pads 18
 – Body more depressed, apterous or, when tegmina present in female never reduced to lateral pads 19
18. Subgenital plate with conical or rounded apex, epiphallus of male genitalia without styli .... *Gryllopsis* Chopard
 – Subgenital plate with apex concave or deeply notched; epiphallus bearing styli on posterolateral corners .... *Turanogryllus* Tarbinskii
19. Completely apterous; anterior tibiae without tympanum .... *Cophogryllus* Saussure
 – Tegmina well developed, at least in male sex; anterior tibiae with at least one tympanum 20
20. Male tegmina with confused mirror and reduced apical field 21
 – Male tegmina with distinct mirror and reduced apical field 22
21. Spines on posterior tibiae feebly movable........................ .... *Nemobiodes* Chopard
 – Spines on posterior tibiae immovable .... *Callogryllus* Sjostedt

22. Epiphallus deeply emarginated posteriorly forming H-shaped structure 23
 – Epiphallus not H-shaped; may be trilobed 24
23. Endoparameres with a broad transverse parameral muscle apodeme, but not with expanded parameral muscle insertion area .... *Tratarogryllus* Tarbinskii
 – Endoparameres with an expanded parameral muscle insertion, but with transverse muscle apodeme .... *Modicogryllus* Chopard
24. Anterior projection of ectoparameres approximately equal 25
 – Internal anterior projection of ectoparameres much longer than external anterior one 26
25. Endoparameres with a transverse parameral muscle apodeme; apex of ectoparamere not digitiform .... *Melanogryllus* Chopard
 – Endoparameres without a transverse parameral muscle apodeme; apex of ectoparamere not digitiform. .... *Gryllus* Linnaeus

GENUS *GRYLLUS* LINNAEUS

The genus *Gryllus* was erected by Linnaeus in 1758 with the help of type species: *Gryllus* (*Acheta*) *campestris* Linnaeus. The genus is cosmopolitan in distribution. Chopard (1969) has placed the genera *Acheta* Fabricius and *Tartarogryllus* Tarbinskii under the genus *Gryllus* Linnaeus. All the three above mentioned genera are very different from one another, chiefly in their male genitalia. Chopard (1969) described 33 species under the genus *Gryllus* from India.

KEY TO THE SPECIES OF GRYLLUS

1. Body and legs almost glabrous. Large species typically black with a yellow spot at base of each tegmina .... *bimaculatus* De Geer
 – Mirror broadly triangular with middle veins reduced, tibia yellowish brown with black spots .... *indica* sp. nov.
 – Body and legs more or less pubescent; colouration different 2

2. Larger species (length of body 18-26 mm) 3
– Smaller species (length of body 8-16 mm) 14
3. Black with four whitish spots on the tegmina .... *quadrimaculatus* Saussure
– Tegmina without whitish spots 4
4. Testaceous; head with a wide transverse brown band; body somewhat depressed .... *domesticus* Linne
– General colour brownish or blackish; body more convex 5
5. Head without any marking 6
– Head presenting a yellow spot or band along each eye 7
6. Larger species (length of body 21-28 mm); pubescent .... *testaceus* Walker
– Smaller species (length of body 18-22 mm); almost glabrous, shining .... *oceanicus* Le Guillou
– Subcosta bearing 12 branches, oblique veins 5 in number, tibiae armed with 9 spines on each superior margin .... *ovali* sp. nov.
7. Tegmina shorter than the abdomen 8
– Tegmina extending to the apex of abdomen 11
8. Very larger species with relatively short posterior femora .... *validus* Chopard
– Smaller species with longer posterior femora 9
9. Tegmina almost black with a very conspicuous ... yellow humeral band (♀) .... *flavovittatus* Chopard
– Tegmina brown 10
10. Mirror a little wider than long, almost rounded .... *himalayanus* Chopard
– Mirror much wider than long, triangular .... *triangulifer* Chopard
11. Head and pronotum blackish, contrasting with the testaceous tegmina; mirror of the male almost apical, non-divided .... *carmichaeli* Chopard
– Tegmina of the same colour as the anterior part of the body 12

12. Posterior legs remarkably long; posterior femora relatively slender .... *gracilipes* Saussure
  - No yellowish spot between the ocelli and eyes, mirror is pentagonal, subcosta bearing 8 branches .... *pentagoni* sp. nov.
  - Posterior legs shorter; posterior femora stouter 13
13. Head presenting two very neat, narrow, yellow bands along the internal margin of the eyes; genitalia of male long, feebly trilobate at apex .... *mitratus* Burmeister
  - Head with a very faint yellow band along the internal margin of the eyes; genitalia short trilobate .... *facialis* Walker
14. Body stout and somewhat depressed 15
  - Body more elongate and convex 16
15. Ovipositor of normal length ....*guttiventris* Walker
  - Ovipositor very short (5.5-6 mm) .... *brunneri* Saussure
16. Head black or brown, without any marking or with very short lines on the occiput 17
  - Head adorned with light bands on the occiput or between the ocelli 24
17. Smaller species (length of body 10-11mm); blackish with fulvous legs 18
  - Legs of the same general colour as the body 19
18. Pronotum with parallel sides; lateral field of tegmina with widely separated veins .... *blennus* Saussure
  - Pronotum slightly enlarged in front; lateral field of tegmina with regularly spaced veins .... *pallipes* Chopard
19. Posterior femora relatively long and slender 20
  - Posterior femora stouter 22
20. Larger species (length of body 15-16 mm); posterior tibiae as long as the femora 21
  - Smaller species (length of body 11 mm); posterior tibiae shorter than the femora ....*fletcheri* Chopard
21. Legs light brown; mirror of male a little longer than broad; 4 oblique veins .... *longipennis* Saussure

– Legs rufous, all the femora blackish at apex; mirror of male as long as broad; 3 oblique veins ...... *bicoloripes* Chopard

22. Colouration uniformly brown or black 23

– Colouration brownish; head black with a yellow spot along the eyes .... *angustulus* Walker

23. General colouration blackish; head black, shining .... *melanocephalus* Serville

– Uniformly chestnut-brown ....*castaneus* Chopard

24. Head with a transverse yellow band between the ocelli 27

– Head without transverse light band 25

25. Head yellowish with three broad brown bands uniting on the rostrum .... *signifrons* Walker

– Head adorned on the occiput with five light bands 26

26. Very smaller species (length of body 8.5 mm); tegmina of male with 4 oblique veins .... *minimus* Chopard

– Larger species (length of body 12 mm); ovipositor very short .... *brevecaudatus* Chopard

27. Colouration uniformly light fulvous brown; pattern of the head feebly marked ....*flavus* Chopard

– Colouration varied with brown or blackish 28

28. Blackish 29

– Brown or testaceous varied with brown 30

29. Head black with a narrow luteous band .... *signipes* Walker

– Anterior femora with yellowish marking above; blackish streaks on the hind femora, tegmina is blackish and not covering the tip of abdomen .... *lanceolati* sp. nov.

– Head blackish with two transverse yellowish bands .... *rehni* Chopard

30. Veins of the lateral field of tegmina regularly and feebly spaced 31

– Veins of the lateral field of tegmina distant and strongly curved ....*confirmatus* Walker

– Tegminal mirror quadrangular, tibia with 8 spines on each superior margin .... *mirroriquadri* sp. nov.

Gryllus ovali sp. nov.

Male (Figure 51)

15.00 mm long excluding anal appendages, body usually more pubescent, more depressed, dark brown to blackish, head 3.00 mm long, thorax 5.00 mm long, tegmina 9.00 mm long, hind wing 17.00 mm long; abdomen 7.00 mm long; anal cerci 7.00 mm long, dark brown, densely hairy.

Head

3.00 mm long, 4.00 mm broad, oval; occiput dark brown to blackish; compound eyes faint brown, ocular distance 0.90 mm; interocular distance 3.00 mm; 3 ocelli arranged in triangle, yellowish brown, circular; ocellar distance 0.18 mm, interocellar distance 0.89 mm; antennae on frontal rostrum; width of frontal rostrum always about double that of first antennal segment; face hairy, face of male convex, frontal rostrum never with prolongation or process; clypeus black, triangular, 0.82 mm long, 1.23 mm broad; labrum dark brown 1.18 mm long, 2.00 mm broad; maxillary palpi yellowish to dark brown, cylindrical, 3.00 mm long, 0.25 mm broad, 5 segmented.

Antenna (Figure 52)

22.00 mm long, 0.80 mm broad, reddish brown; scape 0.50 mm long, 0.53 mm broad; pedicel 0.18 mm long, 0.20 mm broad; flagellum 165 segmented, 21.32 mm long, 0.14 mm broad; 1st segment 0.11 mm long, 0.14 mm broad; 15th segment 0.10 mm long, 0.15 mm broad; terminal segment 0.09 mm long, 0.05 mm broad.

Flagellar Formula

1 = L/W = 0.79, 15 = L/W = 0.67, Ts = L/W = 1.80, A = 1.08

Thorax

5.00 mm long, 4.00 mm broad, yellowish to dark brown, densely hairy, pronotal plate flat.

Tegmina (Figure 53)

Tegmina well developed, long, narrow, extended to 9th abdominal tergite, 9.00 mm long, 5.00 mm broad, yellowish brown, mirror oval, subcosta bearing 12 branches, oblique veins 5 in number.

Hind Wing (Figure 54)

17.00 mm long, 9.00 mm broad, yellowish brown, wings longer than tegmina.

Legs

Fore leg, mid leg lighter than body colour, hind leg darker than body colour.

Fore Leg

9.95 mm long, 1.18 mm broad; coxa 1.30 mm long, 1.18 mm broad, dark brown, densely hairy at outer margin; trochanter 0.65 mm long, 0.54 mm broad, dark brown, sparsely hairy; femur 3.00 mm long, 1.00 mm broad, dark brown, densely hairy; tibia 3.00 mm long, 0.49 mm broad, yellowish to dark brown, anterior tibiae with one tympanum, 3 dark brown tibial spines, densely hairy; tarsus 3 segmented, 2.00 mm long, 0.27 mm broad, dark brown, densely hairy, dark brown tarsal denticles in a row; 1st tarsal segment 1.32 mm long, 0.61 mm broad; 2nd tarsal segment 0.45 mm long, 0.30 mm broad; 3rd tarsal segment 0.55 mm long, 0.18 mm broad; claw dark brown.

Mid Leg

12.45 mm long, 1.09 mm broad; coxa 0.91 mm long, 1.07 mm broad, dark brown, densely hairy; trochanter 0.54 mm long, 0.67 mm broad, yellowish brown, densely hairy; femur 4.00 mm long, 1.00 mm broad, dark brown, densely hairy; tibia 4.00 mm long, 0.47 mm broad, dark brown, densely hairy, 3 dark brown tibial spines; tarsus 3 segmented, 3.00 mm long, 0.32 mm broad, dark brown, densely hairy, reddish brown tarsal denticles in a row; 1st tarsal segment 1.05 mm long, 0.19 mm broad; 2nd tarsal segment 0.40 mm long, 0.27 mm broad; 3rd tarsal segment 1.25 mm long, 0.32 mm broad; claw yellowish brown.

Hind Leg

21.35 mm long, 1.00 mm broad; coxa 2.00 mm long, 1.00 mm broad, yellowish brown, densely hairy at outer margin; trochanter 1.35 mm long, 1.38 mm broad, yellowish brown, sparsely hairy; femur 9.00 mm long, 3.00 mm broad, yellowish to dark brown, densely hairy, length of posterior femora never greater than combined length of tibia and tarsus; tibia 5.00 mm long, 0.66 mm broad, yellowish to

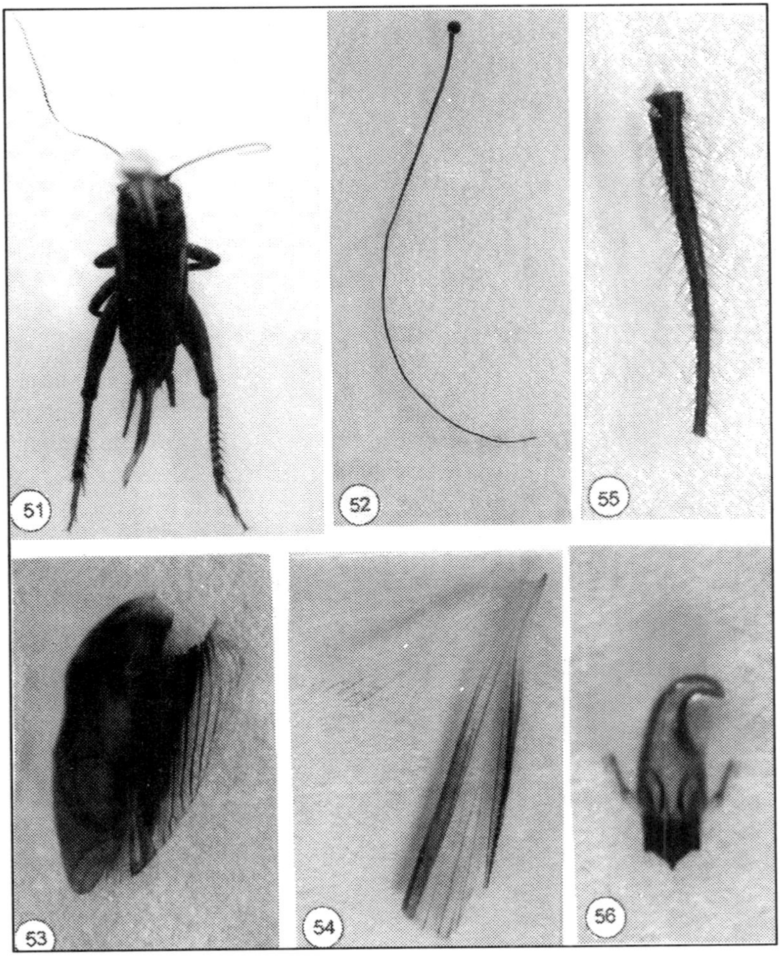

Plate 9: *Gryllus ovali* sp.nov.
Figure 51: Adult: Male; Figure 52: Antenna; Figure 53: Tegmina; Figure 54: Hind Wing; Figure 55: Anal Cerci; Figure 56: Genitalia

dark brown, densely hairy, posterior tibiae without serrulation proximal to spines, 18 reddish brown tibial spines; tarsus 3 segmented, 4.00 mm long, 0.63 mm broad, dark brown, yellowish brown tarsal denticles in a row; 1st tarsal segment 2.00 mm long, 0.30 mm broad; 2nd tarsal segment 0.39 mm long, 0.44 mm broad, minute and compressed; 3rd tarsal segment 1.27 mm long, 0.26 mm broad; claw dark brown.

Abdomen

7.00 mm long, 3.00 mm broad, dark brown to blackish, tapering to posterior side, 10 segmented; anal cerci (Figure 55) 7.00 mm long, dark brown, densely hairy; supra anal plate do not bear a prominent pair of posteriorly projecting heavily sclerotised structure.

Genitalia (Figure 56 and 57)

Genitalia trilobed, yellowish brown; epiphallus not 'H' shaped but trilobed; anterior projections of ectoparameres approximately equal; endoparameres without a transverse paramental muscle apodem, apex of ectoparameres digitiform; superior part of pseudepiphallus 1.84 mm long, 0.77 mm broad; inferior part of pseudepiphallus 1.13 mm long, 0.13 mm broad; lateral rod 2.66 mm long, 0.17 mm broad.

Colour

Dark Brown to Blackish

Occiput, abdomen.

Black

Clypeus.

Faint Brown

Compound eyes.

Yellowish Brown

Ocelli, tegmina, hind wing, mid leg trochanter, claw, hind leg coxa, trochanter, tarsal denticles, genitalia.

Reddish Brown

Antennae, mid leg tarsal denticles, hind leg tibial spines.

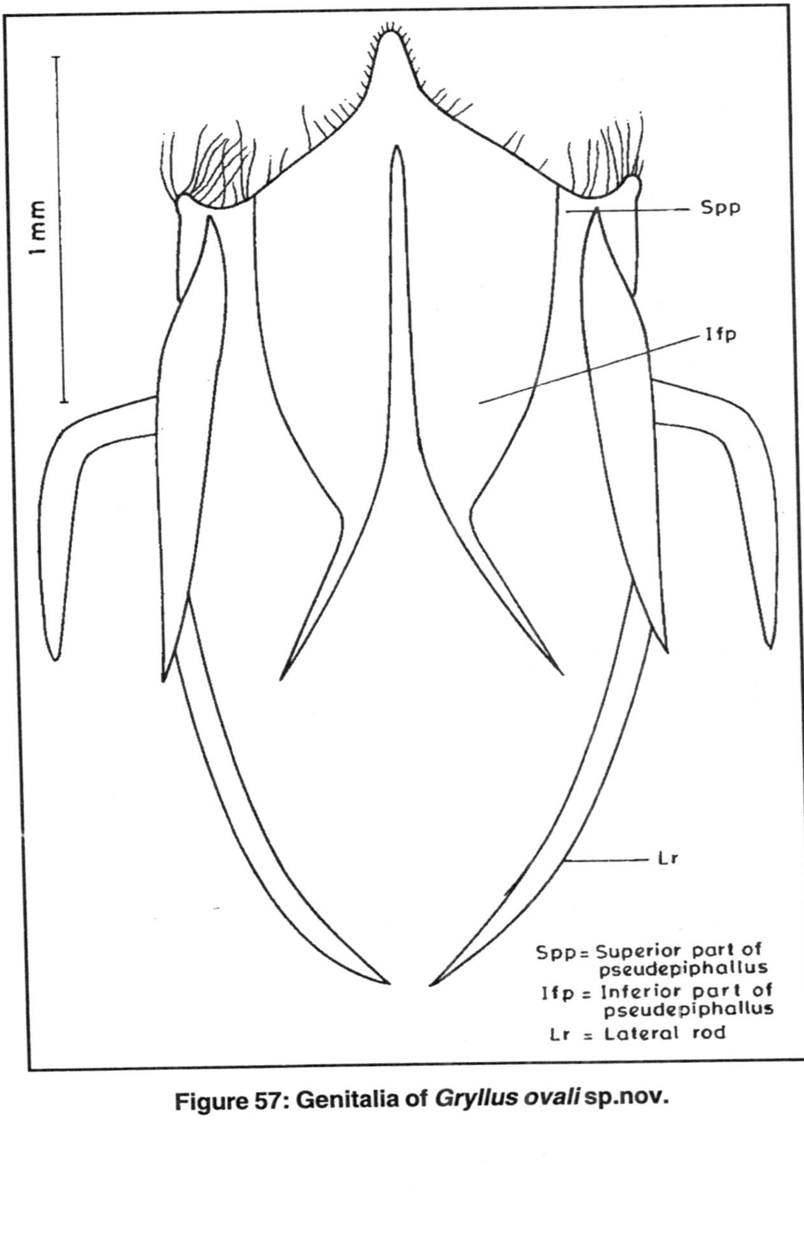

Figure 57: Genitalia of *Gryllus ovali* sp.nov.

Yellowish to Dark Brown

Maxillary palpi, thorax, fore leg tibia, hind leg femur, tibia.

Dark Brown

Labrum, fore leg coxa, trochanter, femur, tibial spines, tarsus, claw, mid leg coxa, femur, tibia, tibial spines, tarsus hind leg tarsus, claw, anal cerci.

Holotype

Male, India, Maharashtra, Coll. 28-XI-2007; Awate Manisha; antenna, wings, cerci, genitalia, on slide labeled as above; other parts with pinned form, labeled as above.

Paratype

4 ♂, 3 ♀, Coll. from 3-VI-2006 to 30-XI-2007 same data as above.

Distributional Record

3 ♂ Kagal, 29-V-2006; 2♂ Chandgad, 5-XI-2006; 1♀ Kolhapur, 5-XI-2006; 1♂ Kolhapur, 31-XII-2006; 1♀ Radhanagari, 3-VI-2007.

Etymology

The mirror on the tegmina in the species is oval hence the name *Gryllus ovali* sp. nov.

Remarks

According to the key of Chopard (1969) and Vasanth (1993) *Gryllus ovali* sp. nov. runs close to *Gryllus oceanicus* Le-guillou by having following characters.

1. Colouration brownish to blackish.
2. Head and pronotum shining.
3. Mirror well developed.

However, it differs from the above species by having following characters.

1. Tibiae armed with 9 spines on each superior margin instead of 6-7.
2. Subcosta bearing 12 branches instead of 5.
3. Mirror shape proportion.
4. Genitalia trilobed, epiphallus not 'H' shaped.

5. Oblique veins 5 instead of 3.
6. Flagellar formula

   1 = L/W = 0.79,15 = L/W = 0.67, Ts = L/W = 1.80, A = 1.08

Gryllus mirroriquadri sp.nov.

Male (Figure 58)

Body pubescent, yellowish to brown, more depressed, more elongated and convex, 16.00 mm long excluding anal appendages, head 3.00 mm long, thorax 6.00 mm long, tegmina 9.00 mm long, hind wing 17.00 mm long; abdomen 7.00 mm long, dark brown to blackish; anal cerci 5.00 mm long, dark brown to blackish, densely hairy.

Head

Head brown with narrow yellow band, 3.00 mm long, 4.00 mm broad, oval; occiput dark brown; compound eyes faint brown, encircled by yellow band, ocular distance 1.41 mm, interocular distance 2.00 mm; head with a transverse yellow band between the ocelli, 3 ocelli intriangle, yellowish brown, circular, ocellar distance 0.25 mm, interocellar distance 1.05 mm; antennae in frontal rostrum, face of the male convex, frontal rostrum never with prolongation, frontal rostrum always about double that of first antennal segment; clypeus dark brown, triangular, 0.91 mm long, 0.93 mm broad; labrum yellowish brown 0.83 mm long, 1.29 mm broad; maxillary palpi dark brown, cylindrical, 3.00 mm long, 0.20 mm broad, 5 segmented.

Antenna (Figure 59)

13.00 mm long, 0.55 mm broad, yellowish brown; scape 0.45 mm long, 0.55 mm broad; pedicel 0.20 mm long, 0.17 mm broad; flagellum 108 segmented, 12.35 mm long, 0.15 mm broad; 1st segment 0.08 mm long, 0.15 mm broad; 15th segment 0.09 mm long, 0.11 mm broad; terminal segment 0.07 mm long, 0.08 mm broad.

Flagellar Formula

1 = L/W = 0.53,15 = L/W = 0.81, Ts = L/W = 0.87, A = 0.73

Thorax

6.00 mm long, 4.00 mm broad, dark brown, densely hairy, pronotum with yellow broken convex band, lateral lobes rectangular.

Tegmina (Figure 60)

Tegmina well developed, long, narrow, extended to 9th abdominal tergite, 9.00 mm long, 4.00 mm broad, yellowish brown; male tegmina with distinct mirror and apical field, mirror longer than wide, oblique veins 3 in number, apical field with 7 veins.

Hind Wing (Figure 61)

17.00 mm long, 5.00 mm broad, yellowish brown, wings longer than tegmina.

Legs

Fore leg, mid leg, hind leg lighter than body colour.

Fore Leg

8.47 mm long, 1.04 mm broad; coxa 0.70 mm long, 1.04 mm broad, dark brown, densely hairy; trochanter 0.77 mm long, 0.72 mm broad, dark brown, sparsely hairy; femur 3.00 mm long, 1.00 mm broad, yellowish brown, densely hairy; tibia 2.00 mm long, 0.36 mm broad, dark brown, 3 dark brown tibial spines, densely hairy, anterior with one tympanum; tarsus 3 segmented, 2.00 mm long, 0.22 mm broad, dark brown, densely hairy, dark brown tarsal denticles in a row; 1st tarsal segment 0.79 mm long, 0.20 mm broad; 2nd tarsal segment 0.12 mm long, 0.20 mm broad; 3rd tarsal segment 0.83 mm long, 0.16 mm broad; claw yellowish to dark brown.

Mid Leg

9.19 mm long, 1.29 mm broad; coxa 1.12 mm long, 1.29 mm broad, dark brown, densely hairy; trochanter 0.91 mm long, 1.04 mm broad, dark brown, sparcely hairy; femur 3.00 mm long, 0.83 mm broad, yellowish to dark brown, densely hairy; tibia 2.00 mm long, 0.37 mm broad, yellowish brown, densely hairy, 3 dark brown tibial spines; tarsus 3 segmented, 2.16 mm long, 0.25 mm broad, dark brown, densely hairy; 1st tarsal segment 2.00 mm long, 0.26 mm broad; 2nd tarsal segment 0.28 mm long, 0.20 mm broad; 3rd tarsal segment 0.79 mm long, 0.17 mm broad; claw dark brown.

Hind Leg

20.91 mm long, 2.00 mm broad; coxa 2.00 mm long, 2.00 mm broad, yellowish brown, densely hairy; trochanter 0.91 mm long, 0.83 mm broad, dark brown, sparsely hairy; length of posterior femora

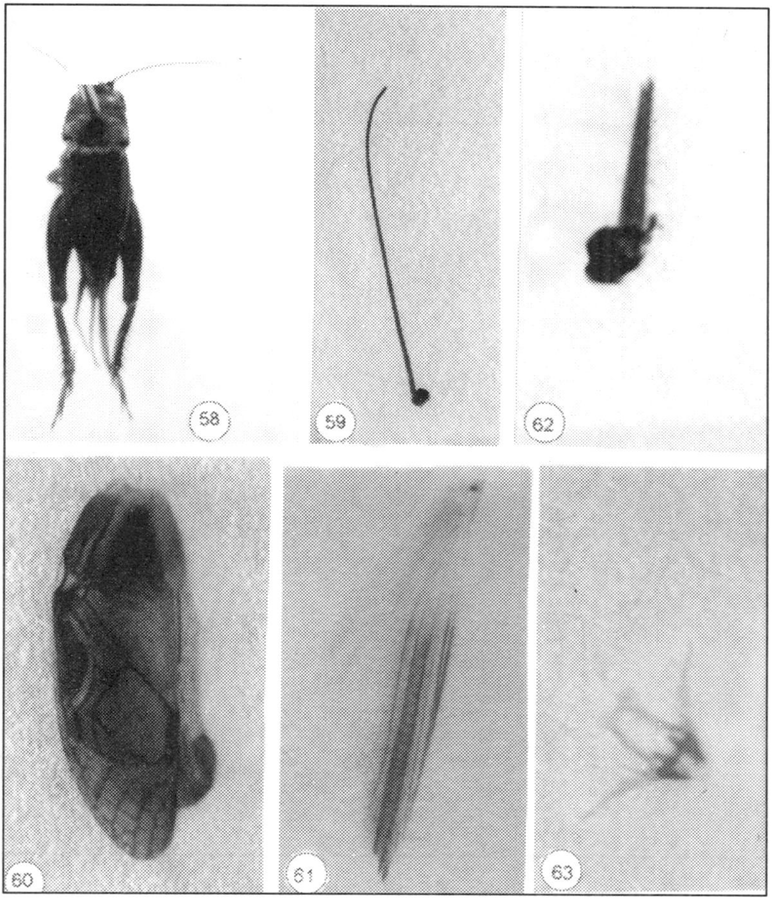

Plate 10: ***Gryllus mirroriquadri*** **sp.nov.**
Figure 58: Adult: Male; Figure 59: Antenna; Figure 60: Tegmina; Figure 61: Hind Wing; Figure 62: Anal Cerci; Figure 63: Genitalia

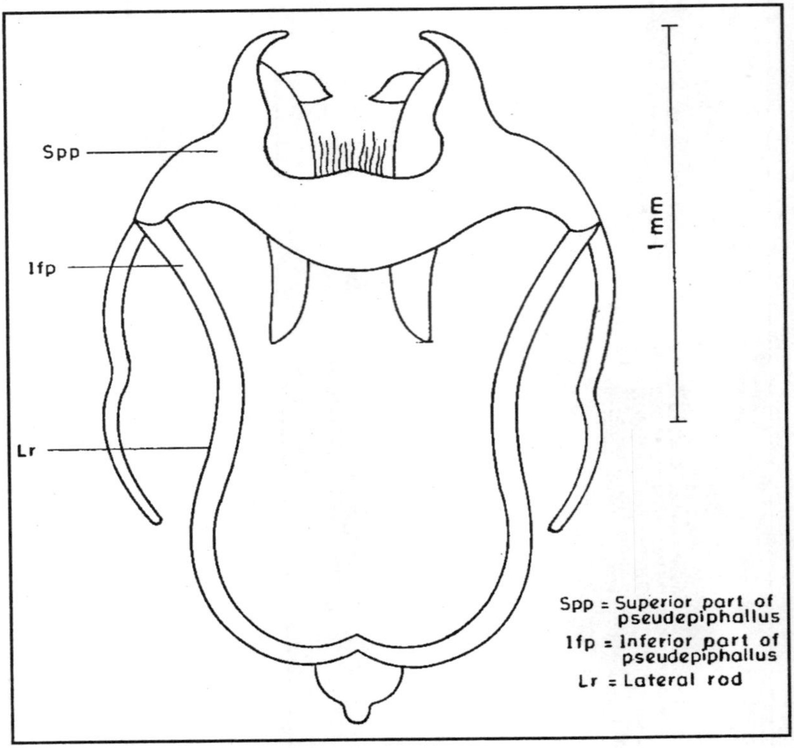

Figure 64: Genitalia of *Gryllotalpa mirroriquadri* sp.nov.

equal but never greater than combined length of tibia and tarsus, femur 8.00 mm long, 3.00 mm broad, dark brown, densely hairy; posterior tibiae without serrulation proximal to spines, tibia 5.00 mm long, 0.30 mm broad, dark brown, densely hairy, 16 yellowish to dark brown tibial spines; tarsus 3 segmented, 5.00 mm long, 0.32 mm broad, dark brown, densely hairy, reddish brown tarsal denticles in a row; 1st tarsal segment 3.00 mm long, 0.31 mm broad; 2nd tarsal segment 0.29 mm long, 0.20 mm broad; 3rd tarsal segment 1.29 mm long, 0.18 mm broad; claw dark brown.

Abdomen

7.00 mm long, 4.00 mm broad, dark brown above and yellow beneath, tapering to posterior side, 10 segmented; anal cerci (Figure 62) 5.00 mm long, dark brown, densely hairy; supra anal plate do not bear a prominent pair of posteriorly projecting heavily sclerotised structure.

Genitalia (Figure 63 and 64)

Superior part of pseudepiphallus reddish brown, 1.3 mm long, 0.16 mm broad, long hairs present; lateral rod 1.41 mm long, 0.16 mm broad; inferior part of pseudepiphallus 0.73 mm long, 0.26 mm broad; genitalia with superior bridge deeply and widely notched, forming two acute, erect lateral lobes; inferior pieces presenting well developed erect tooth.

Colour

Dark Brown

Occiput, maxillary palpi, thorax, fore leg coxa, tibia, tarsus, mid leg coxa, trochanter, tibial spines, tarsus, hind leg trochanter, femur, tibia, tarsus, mid, hind leg claw.

Faint Brown

Compound eyes.

Dark Black

Clypeus.

Dark Brown to Yellow

Abdomen.

Yellowish Brown

Labrum, ocelli, antennae, tegmina, hind wing, fore leg femur, mid leg tibia, hind leg coxa.

Yellowish to Dark Brown

Mid leg femur, hind leg tibial spines, fore leg claw, anal cerci.

Reddish Brown

Genitalia

Holotype

Male, India, Maharashtra, Coll. 29-X-2006; Awate Manisha; antenna, wings, cerci, genitalia, on slide labeled as above other parts with pinned form, labeled as above.

Paratype

5 ♂, 5 ♀, Coll. from 3-VI-2006 to 30-XI-2007 same data as above.

Distributional Record

3 ♂ Panhala, 23-VII-2006; 2♂ Kagal, 29-VI-2006; 3 ♂ Bhudargad, 15–VI-2007; 2 ♀ Kolhapur, 28-VIII-2007.

Etymology

The mirror of the tegmina is quadrangular in the species hence the name *Gryllus mirroriquadri* sp. nov.

Remarks

According to the key of Chopard (1969) and Vasanth (1993) *Gryllus mirroriquadri* sp. nov. runs close to *Gryllus confirmatus* Walker by having following characters.

1. Head brown with narrow yellow band.
2. Abdomen dark brown above.
3. Legs yellow.

However, it differs from above species by having following characters

1. Pronotum with yellow broken convex band.
2. Eyes encircled by yellow band.
3. Antennal length proportion to the body.

4. Tibia with 8 spines on each superior margin instead of 5.
5. Tegminal mirror quadrangular.
6. Apical field with 7 veins instead of 5.
7. Flagellar formula

   $1 = L/W = 0.53, 15 = L/W = 0.81, Ts = L/W = 0.87, A = 0.73$
8. Genitalia having well developed erect tooth and superior part of pseudepiphallus having long hairs.

Gryllus mirroriquadri sp.nov.

Female (Figure 65)

Body yellowish to dark dark brown, pubescent, elongate and convex, more depressed, 18.00 mm long excluding anal appendages, head 4.00 mm long, thorax 6.00 mm long, tegmina 11.00 mm long, hind wing 18.00 mm long; abdomen 8.00 mm long, blackish; anal cerci 6.00 mm long, dark brown, ovipositor 6.00 mm long, yellowish to reddish brown.

Head

4.00 mm long, 5.00 mm broad, oval; occiput brown to blackish; compound eyes faint brown, ocular distance 1.08 mm, interocular distance 3.00 mm; head with yellow band between ocelli, 3 ocelli arranged in triangle, yellowish brown, circular, ocellar distance 0.20 mm, interocellar distance 0.98 mm; antennae on frontal rostrum, face of the female convex, frontal rostrum never with prologation, frontal rostrum always about double that of first antennal segment, hairy; clypeus blackish, oval, 1.66 mm long, 1.34 mm broad; labrum dark brown, semicircular, 0.91 mm long, 2.00 mm broad; maxillary palpi dark brown, cylindrical, 4.00 mm long, 0.28 mm broad, 5 segmented.

Antenna (Figure 66)

16.00 mm long, 0.19 mm broad, reddish brown; scape 0.35 mm long, 0.60 mm broad; pedicel 0.26 mm long, 0.17 mm broad; flagellum 163 segmented, 16.39 mm long, 0.20 mm broad; 1st segment 0.12 mm long, 0.20 mm broad; 15th segment 0.15 mm long, 0.10 mm broad; terminal segment 0.10 mm long, 0.05 mm broad, hairy.

Flagellar Formula

1 = L/W = 0.6, 15 = L/W = 1.5, Ts = L/W = 2, A = 1.36

Thorax

6.00 mm long, 5.00 mm broad, yellowish to dark brown, densely hairy, pronotum tranverse with anterior margin feebly concave.

Tegmina (Figure 67)

Wings in female are always dissimilar with males of the same species, long, narrow, extended to last abdominal tergite, 11.00 mm long, 5.00 mm broad, yellowish brown, 5 branches of cubital, veins of lateral field of tegmina curved less than male tegmina.

Hind Wing (Figure 68)

18.00 mm long, 7.00 mm broad, yellowish brown, wings longer than tegmina.

Legs

Fore leg, mid leg lighter than body colour; hind leg darker than body colour.

Fore Leg

12.15 mm long, 1.08 mm broad, coxa 1.20 mm long, 1.08 mm broad, dark brown, densely hairy at outer margin; trochanter 0.95 mm long, 0.86 mm broad, dark brown, sparsely hairy at outer margin; femur 4.00 mm long, 1.12 mm broad, dark brown, densely hairy; tibia 3.00 mm long, 0.53 mm broad, dark brown, 3 dark brown tibial spines, densely hairy; tarsus 3 segmented, 3.00 mm long, 0.31 mm broad, dark brown, densely hairy, denticles present in rows; 1st tarsal segment 1.37 mm long, 0.22 mm broad; 2nd tarsal segment 0.37 mm long, 0.24 mm broad; 3rd tarsal segment 0.79 mm long, 0.2 mm broad; claw reddish brown.

Mid Leg

10.99 mm long, 1.04 mm broad, coxa 1.16 mm long, 1.04 mm broad, dark brown, densely hairy; trochanter 0.83 mm long, 0.43 mm broad, yellowish brown, sparsely hairy; femur 3.00 mm long, 1.00 mm broad, yellowish brown, densely hairy; tibia 3.00 mm long, 0.42 mm broad, dark brown, densely hairy, dark brown 3 tibial spines;

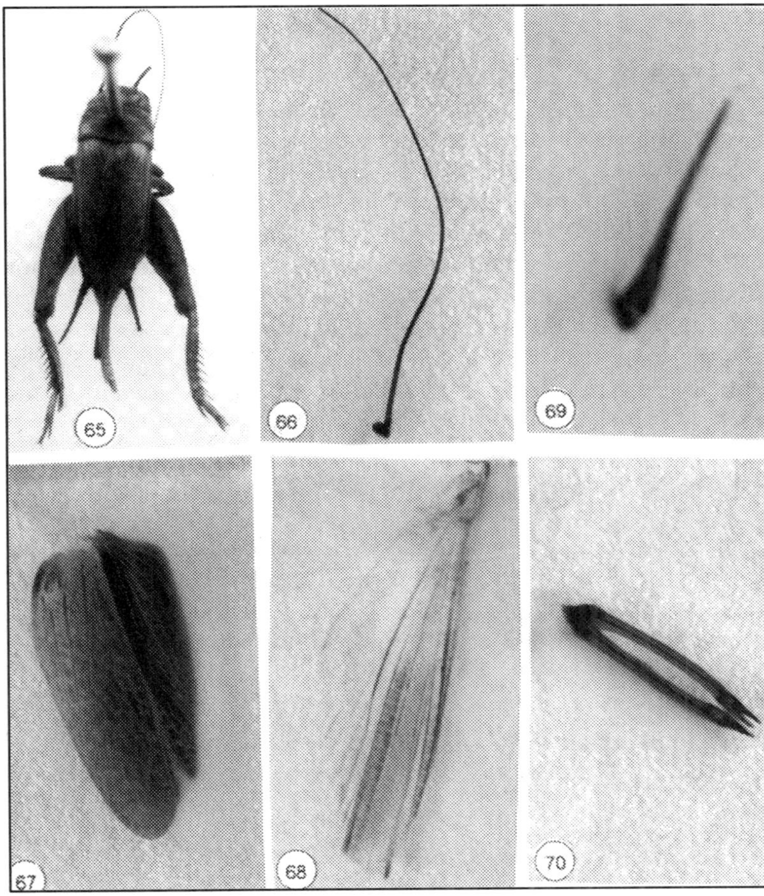

Plate 11: *Gryllus mirroriquadri* sp.nov.
Figure 65: Adult: Female; Figure 66: Antenna; Figure 67: Tegmina; Figure 68: Hind Wing; Figure 69: Anal Cerci; Figure 70: Ovipositor

tarsus 3 segmented, 3.00 mm long, 0.18 mm broad, dark brown, densely hairy, dark brown tarsal denticles present in rows; 1st tarsal segment 2.00 mm long, 0.25 mm broad; 2nd tarsal segment 0.37 mm long, 0.29 mm broad; 3rd tarsal segment 1.33 mm long, 0.37 mm broad; claw reddish brown.

Hind Leg

25.00 mm long, 2.00 mm broad, coxa 2.00 mm long, 2.00 mm broad, yellowish brown, sparsely hairy; trochanter 2.00mm long, 1.92 mm broad, grayish brown, sparsely hairy; length of posterior femora equal but never greater than combined length of tibia and tarsus, femur 10.00 mm long, 4.00 mm broad, yellowish brown to dark brown, densely hairy; posterior tibiae without serrulation proximal to spines, tibia 7.00 mm long, 0.69 mm broad, dark brown, 16 reddish brown tibial spines, densely hairy; tarsus 3 segmented, 4.00 mm long, 0.33 mm broad, dark brown, densely hairy, tarsal denticles present in rows; 1st tarsal segment 2.00 mm long, 0.38 mm broad; 2nd tarsal segment 0.35 mm long, 0.39 mm broad; 3rd tarsal segment 1.37 mm long, 0.21 mm broad; claw reddish brown.

Abdomen

8.00 mm long, 5.00 mm broad, yellowish brown to dark brown, tapering to posterior side, 9 segmented; anal cerci (Figure 69) 6.00 mm long, dark brown; ovipositor (Figures 70 and 71) shorter 6.00 mm long, 0.5 mm broad, acute apex, yellowish to reddish brown.

Colour

Brown to Blackish

Occiput, clypeus.

Faint Brown

Compound eyes.

Yellowish Brown

Ocelli, tegmina, hind wing, mid leg trochanter, hind leg coxa.

Reddish Brown

Antennae, fore leg claw, mid leg tibial spines, claw, hind leg tibial spines, claw.

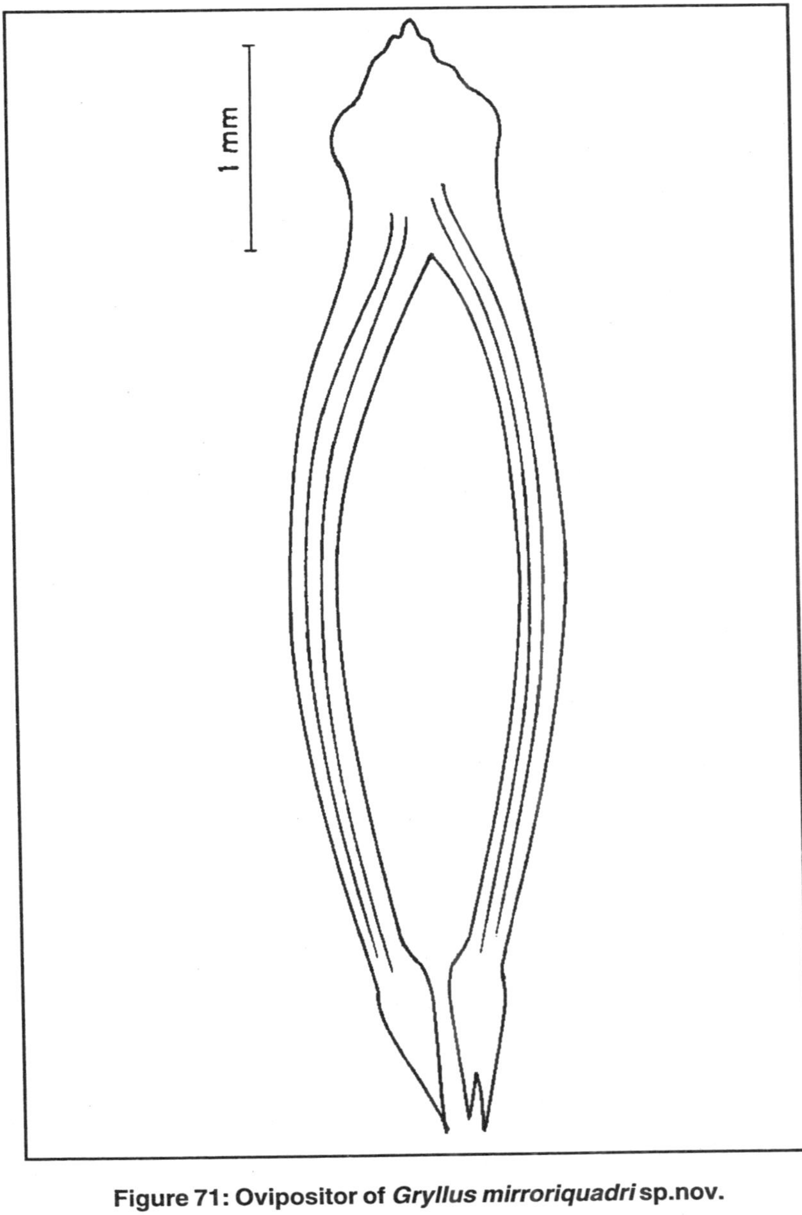

Figure 71: Ovipositor of *Gryllus mirroriquadri* sp.nov.

Yellowish Brown to Dark Brown

Thorax, hind leg femur, abdomen.

Dark Brown

Labrum, maxillary palpi, fore leg coxa, trochanter, femur, tibia, tibial spines, tarsus, mid leg coxa, femur, tibia, tarsus, hind leg tibia, tarsus, anal cerci.

Grayish Brown

Hind leg trochanter.

Yellowish to Reddish Brown

Ovipositor.

Holotype

Female, India, Maharashtra, Coll. 28-X-2006; Awate Manisha; antenna, wings, cerci, ovipositor on slide labeled as above; other parts with pinned form, labeled as above.

Paratype

6 ♂, 4 ♀, Coll. from 1-VI-2006 to 29–XII – 2007, same data as above.

Distributional Record

3♂ Ajara, 3-VI-2006; 2♂ Karveer, 29-VII-2007; 1♀ Gaganbavada, 24-IX-2006; 1♂ Kagal, 16-IX-2007; 2♀ Kagal, 6-IX-2007; 1 ♀ Kolhapur, 7-X-2007.

Etymology

As per given under description of male.

Remarks

According to the key of Chopard (1969) and Vasanth (1993) *Gryllus mirroriquadri* sp. nov. runs close to *Gryllus confirmatus* Walker by having following characters.

1. Head brown with yellow band.
2. Abdomen dark brown above.
3. Legs yellow.

However, it differs from above species by having following characters

1. Antennal length proportion to the body.
2. Five branches of cubital.
3. Tibia with 8 spines on each superior margin instead of 5.
4. Ovipositor short, acute apex.
5. Flagellar formula

   1 = L/W = 0.6, 15 = L/W = 1.5, Ts = L/W = 2, A = 1.36

Gryllus indica sp.nov.

Male (Figure 72)

Body and legs glabrous black with yellow spot at base of each tegmina, body depressed, not so convex, pubescent, 17.00 mm long excluding anal appendages, head 3.00 mm long; thorax 7.00 mm long, tegmina 9.00 mm long, hind wing 16.00 mm long; abdomen 7.00 mm long; anal cerci 5.00 mm long, dark brown, densely hairy.

Head

3.00 mm long, 4.00 mm broad, oval; occiput blackish; compound eyes dark black, ocular distance 1.56 mm, interocular distance 3.00 mm; 3 ocelli arranged in triangle, yellowish brown, circular, ocellar distance 0.34 mm, interocellar distance 0.78 mm; head with wide frontal rostrum, frontal rostrum never with prolongation, width of frontal rostrum always about double that of first antennal segment; antennae on frontal rostrum, face of male convex; clypeus dark brown, quadrangular, 0.79 mm long, 2.00 mm broad; labrum yellowish to dark brown, semicircular, 0.55 mm long, 1.54 mm broad; maxillary palpi yellowish brown, cylindrical, 4.00 mm long, 0.24 mm broad, 5 segmented.

Antenna (Figure 73)

16.00 mm long, 0.46 mm broad, reddish brown; scape 0.48 mm long, 0.46 mm broad; pedicel 0.22 mm long, 0.15 mm broad; flagellum 163 segmented, 15.3 mm long, 0.15 mm broad; 1st segment 0.10 mm long, 0.15 mm broad; 15th segment 0.08 mm long, 0.12 mm broad; terminal segment 0.12 mm long, 0.05 mm broad.

Flagellar Formula

1 = L/W = 0.66, 15 = L/W = 0.66, Ts = L/W = 2.40, A = 1.24

Thorax

7.00 mm long, 5.00 mm broad, yellowish brown to dark brown, densely hairy, pronotum strongly widening.

Tegmina (Figure 74)

Tegmina well developed, long, narrow, not extending to the apex of abdomen, 9.00 mm long, 6.00 mm broad, yellowish to dark brown, male tegmina with distinct mirror and apical field, mirror longer than wide, mirror broadly triangular with middle veins reduced, apical field well developed with reticulation, first chord veinlets absent; several oblique veins, subcosta bearing 11 branches.

Hind Wing (Figure 75)

16.00 mm long, 7.00 mm broad, yellowish brown, wings longer than tegmina.

Legs

Fore leg, mid leg, hind leg lighter than body colour.

Fore Leg

9.14 mm long, 1.09 mm broad; coxa 1.73 mm long, 1.09 mm broad, yellowish to dark brown, densely hairy; trochanter 0.41 mm long, 0.69 mm broad, dark brown, sparsely hairy; femur 3.00 mm long, 0.80 mm broad, yellowish to dark brown, densely hairy ; tibia 2.00 mm long, 0.43 mm broad, yellowish to dark brown, densely hairy, 2 yellowish brown tibial spines; tarsus 3 segmented, 2.00 mm long, 0.24 mm broad, dark brown, densely hairy, dark brown tarsal denticles in a row; 1st tarsal segment 1.08 mm long, 0.37 mm broad; 2nd tarsal segment 0.6 mm long, 0.25 mm broad; 3rd tarsal segment 0.25 mm long, 0.19 mm broad; claw dark brown.

Mid Leg

10.24 mm long, 0.86 mm broad; coxa 0.76 mm long, 0.86 mm broad, yellowish to dark brown, sparsely hairy at outer margin; trochanter 0.48 mm long, 0.66 mm broad, dark brown, sparsely hairy; femur 3.00 mm long, 0.83 mm broad, dark brown, densely hairy;

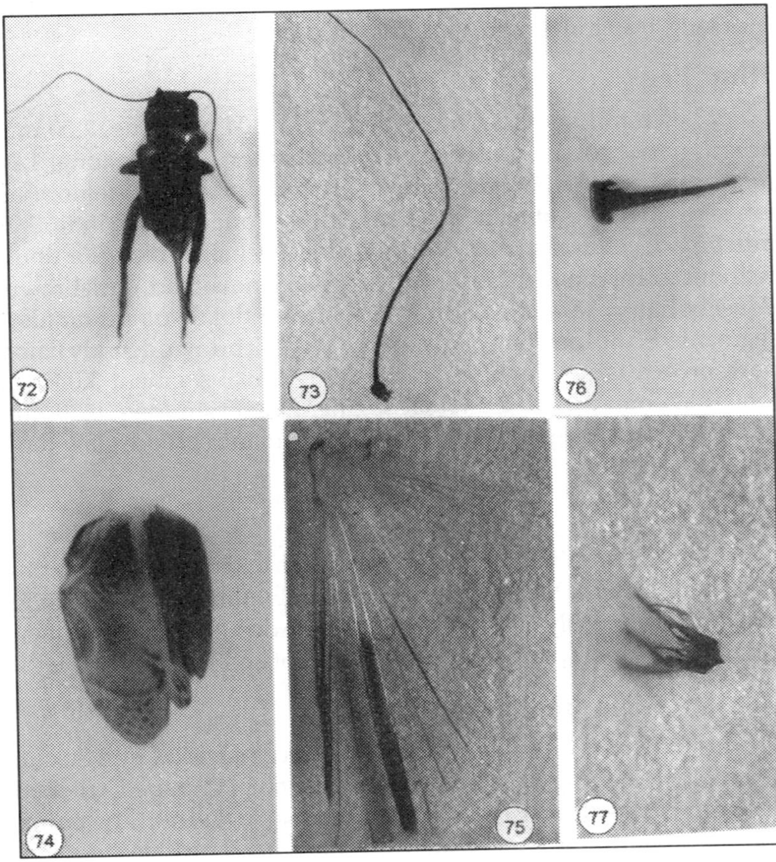

Plate 12: *Gryllus indica* sp.nov.
Figure 72: Adult: Male; Figure 73: Antenna; Figure 74: Tegmina; Figure 75: Hind Wing; Figure 76: Anal Cerci; Figure 77: Genitalia

tibia 3.00 mm long, 0.41 mm broad, dark brown, densely hairy, 3 dark brown tibial spines; tarsus 3 segmented, 3.00 mm long, 0.27 mm broad, dark brown, densely hairy, dark brown tarsal denticles in a row; 1st tarsal segment 1.16 mm long, 0.28 mm broad; 2nd tarsal segment 0.20 mm long, 0.19 mm broad; 3rd tarsal segment 1.04 mm long, 0.15 mm broad; claw dark brown.

Hind Leg

23.04 mm long, 2.00 mm broad; coxa 2.04 mm long, 2.00 mm broad, dark brown, densely hairy; trochanter 2.00 mm long, 1.00 mm broad, yellowish brown, sparsely hairy; posterior femora no longer than tibia and metatarsus together, femur 8.00 mm long, 3.00 mm broad, yellowish to dark brown, densely hairy; posterior tibiae without serrulation, tibia 6.00 mm long, 0.55 mm broad, dark brown, densely hairy, 18 yellowish to dark brown tibial spines; tarsus 3 segmented, 5.00 mm long, 0.33 mm broad, dark brown, densely hairy, dark brown tarsal denticles in a row; 1st tarsal segment 3.00 mm long, 0.33 mm broad; 2nd tarsal segment 0.66 mm long, 0.20 mm broad; 3rd tarsal segment 1.16 mm long, 0.16 mm broad; claw dark brown.

Abdomen

7.00 mm long, 3.00 mm broad, yellowish to dark brown, tapering to posterior side, 10 segmented; anal cerci (Figure 76) 5.00 mm long, dark brown, densely hairy; supra anal plate do not bear a prominent pair of posteriorly projecting heavily sclerotised structure.

Genitalia (Figures 77 and 78)

Genitalia plain, formed of a large pieces divided into 3 acute apex, 2 lobes having long hairs; epiphallus not 'H' shaped, trilobed; anterior projection of ectoparameres approximately equal; endoparameres without a transverse parameral muscle; apodeme apex of ectoparamere digitiform; superior part of pseudepiphallus 0.75 mm long, 0.1 mm broad; lateral rod 1.3 mm long, 0.08 mm broad; inferior part of pseudepiphallus 0.71 mm long, 0.28 mm broad, yellowish brown.

Colour

Blackish

Occiput.

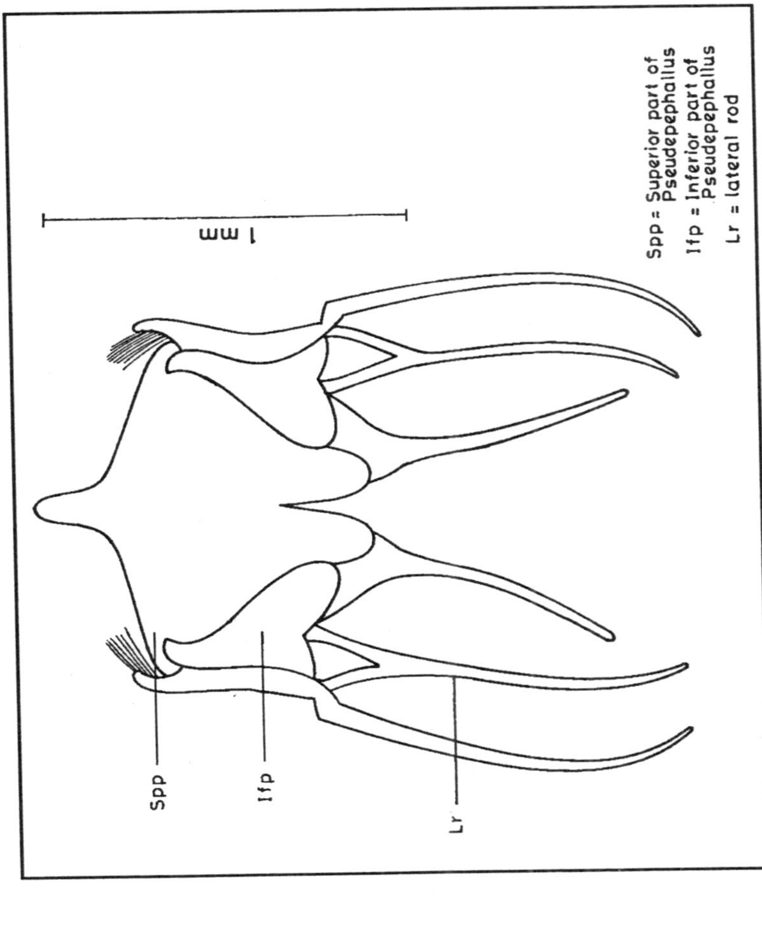

Figure 78: Genitalia of *Gryllotalpa indica* sp.nov.

Dark Black

Compound eyes.

Yellowish Brown

Maxillary palpi, ocelli, hind wing, fore leg tibial spines, hind leg trochanter.

Reddish Brown

Antennae.

Yellowish Brown to Dark Brown

Labrum, thorax, tegmina, fore leg coxa, femur, tibia, mid leg coxa, hind leg femur, abdomen.

Dark Brown

Clypeus, fore leg trochanter, tarsus, tarsal denticles, claws, mid leg trochanter, femur, tibia, tibial spines, tarsus, tarsal denticles, claw, anal cerci.

Holotype

Male, India, Maharashtra, Coll.11-X-2006; Awate Manisha; antenna, wings, genitalia on slide labeled as above; other parts with pinned form, labeled as above.

Paratype

5 ♂, 3 ♀, Coll. from 2-VII-2006 to 30-XI-2007, same data as above.

Distributional Record

1♂ Hatkangale, 16-VII-2006; 1♂ Kolhapur, 26-VIII-2007; 1♂ Bhudargad 15-VI-2007; 1♂ Karveer, 2-IX-2007; 1♂ Ajara, 30-IX-2007; 2♀ Ajara 30-IX-2007; 1♀ Kagal, 6-X-2007.

Etymology

Gryllus indica sp. nov. has been reported from India for the first time, hence the name.

Remarks

According to the key of Chopard (1969) and Vasanth (1993) *Gryllus indica* sp. nov. runs close to *Gryllus bimaculatus* De Geer by having following characters.

1. Body glaborus black colour.
2. Yellow spot at base of tegmina.

However, it differs from the above species by having following characters.

1. Tegmina not extending to the apex of abdomen.
2. Posterior tibia armed with 9 spines on each superior margin.
3. Mirror broadly triangular with middle veins reduced.
4. Apical field well developed with reticulation.
5. Subcosta bearing 11 branches.
6. Oblique veins 4 in number.
7. Flagellar formula

   $1 = L/W = 0.66, 15 = L/W = 0.66, Ts = L/W = 2.40, A = 1.24$

Gryllus indica sp.nov.

Female (Figure 79)

Body and legs glaborous black yellow spot at base, less convex, pubescent, 16.00 mm long excluding anal appendages, head 3.00 mm long, thorax 6.00 mm long, tegmina 10.00 mm long, hind wing 18.00 mm long; abdomen 7.00 mm long, dark brown; anal cerci 6.00 mm long, dark brown; ovipositor 7.00 mm long, reddish brown.

Head

3.00 mm long, 4.00 mm broad; rounded; occiput dark brown to blackish; compound eyes black; ocular distance 1.04 mm, interocular distance 3.00 mm; 3 ocelli disposed in triangle, reddish brown, circular, ocellar distance 0.34 mm, interocellar distance 1.04 mm; head with wide frontal rostrum, frontal rostrum never with prolongation, width of frontal rostrum always about double that of first antennal segment; antennae on frontal rostrum, face convex; clypeus, dark brown, semicircular, 0.37 mm long, 1.20 mm broad; labrum, yellowish brown, semicircular, 0.85 mm long, 1.41 mm broad; maxillary palpi, yellowish brown, cylindrical, 4.03 mm long, 0.30 mm broad, 5 segmented.

Antenna (Figure 80)

20.00 mm long, 0.45 mm broad, yellowish to dark brown, antennal margin dark brown; scape 0.49 mm long, 0.59 mm broad, hairy; pedicel 0.22 mm long, 0.19 mm broad; flagellum 142 segmented, 19.29 mm long, 0.18 mm broad; 1st segment 0.11 mm long, 0.18 mm broad; 15th segment 0.09 mm long, 0.13 mm broad; terminal segment 0.12 mm long, 0.05 mm broad.

Flagellar Formula

1 = L/W = 0.61, 15 = L/W = 0.69, Ts = L/W = 2.30, A = 1.20

Thorax

6.00 mm long, 5.00 mm broad, dark brown, sparsely hairy, pronotum superior, flat.

Tegmina (Figure 81)

Tegmina well developed never reduced to small lateral pads, long, narrow, extended to last abdominal tergite, 10.00 mm long, 5.00 mm broad, dark brown, sparsely hairy.

Hind Wing (Figure 82)

18.00 mm long, 3.00 mm broad, yellowish brown, wings longer than tegmina.

Legs

Legs colour similar to body colour.

Fore Leg

11.38 mm long, 2.00 mm broad; coxa 1.00 mm long, 2.00 mm broad, dark brown, densely hairy; trochanter 0.95 mm long, 0.93 mm broad, dark brown, densely hairy; femur 4.00 mm long, 1.07 mm broad, dark brown to blackish, densely hairy; tibia 2.43 mm long, 0.55 mm broad, dark brown, two dark brown tibial spines at the tip; tarsus 3 segmented, 3.00 mm long, 0.26 mm broad, dark brown; 1st tarsal segment 0.75 mm long, 0.26 mm broad; 2nd tarsal segment 0.25 mm long, 0.26 mm broad; 3rd tarsal segment 1.16 mm long, 0.20 mm broad; claw dark brown.

Mid Leg

12.26 mm long, 0.60 mm broad; coxa 2.00 mm long, 0.60 mm broad, dark brown to blackish, hairy; trochanter 0.86 mm long, 0.80

Plate 13: *Gryllus indica* sp.nov.
Figure 79: Adult: Female; Figure 80: Antenna; Figure 81: Tegmina; Figure 82: Hind Wing; Figure 83: Anal Cerci; Figure 84: Ovipositor

mm broad, dark brown, densely hairy; femur 4.00 mm long, 0.90 mm broad, dark brown to blackish, densely hairy; tibia 3.00 mm long, 0.48 mm broad, dark brown to blackish, densely hairy, 3 tibial spines; tarsus 3 segmented, 2.4 mm long, 0.22 mm broad, dark brown to blackish, sparsely hairy; 1st tarsal segment 1.29 mm long, 0.28 mm broad; 2nd tarsal segment 0.25 mm long, 0.33 mm broad; 3rd tarsal segment 1.33 mm long, 0.25 mm broad; claw reddish brown.

Hind Leg

22.86 mm long, 3.00 mm broad; coxa 2.00 mm long, 3.00 mm broad, brown to blackish; trochanter 0.86 mm long, 1.56 mm broad, dark brown to blackish, hairy; posterior femora no longer than tibia and metatarsus together, femur 9.00 mm long, 3.00 mm broad, yellowish brown to dark brown, densely hairy; posterior tibia strong brownish black, none serrulated at base, tibia 6.00 mm long, 1.00 mm broad, tibia yellowish brown with black spots, 16 reddish brown tibial spines, densely hairy; tarsus 3 segmented, 5.00 mm long, 0.40 mm broad, dark brown, sparsely hairy, 8 reddish brown tarsal spine; 1st tarsal segment 2.00 mm long, 0.42 mm broad; 2nd tarsal segment 1.25 mm long, 0.25 mm broad; 3rd tarsal segment 1.75 mm long, 0.39 mm broad; claw reddish brown.

Abdomen

7.00 mm long, 4.00 mm broad, tapering to posterior side, yellowish brown to dark brown banded tergite, 10 segmented; anal cerci (Figure 83) 6.00 mm long, dark brown, hairy; ovipositor (Figures 84 and 85) 7.00 mm long, reddish brown, ovipositor shorter than posterior femora with apical valve lanceolate.

Colour

Dark Brown to Blackish

Occiput, fore leg femur, tibia, tarsus, claw mid leg coxa, femur, tibia, tarsus, hind leg coxa, trochanter.

Black

Compound eyes.

Reddish Brown

Ocelli, hind leg tibial, tarsal spines, tarsus, claw, ovipositor.

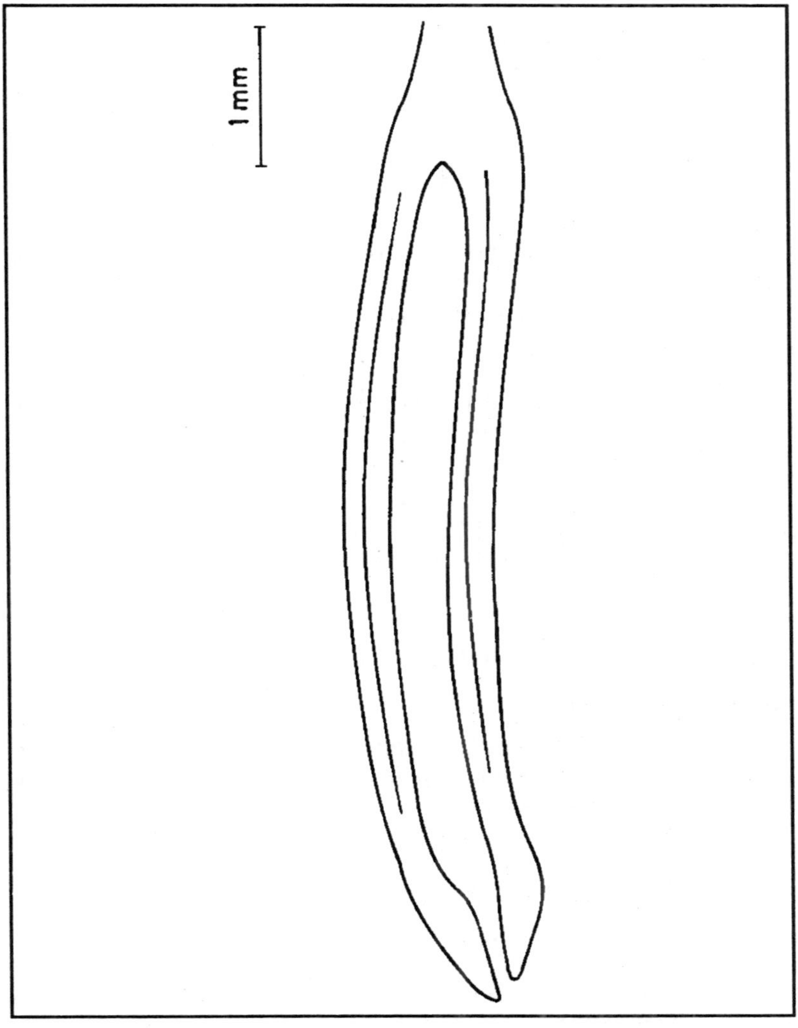

Figure 85: Ovipositor of *Gryllus indica* sp.nov.

Yellowish to Dark Brown

Antennae, hind leg femur, abdomen.

Dark Brown

Clypeus, thorax, tegmina, fore leg coxa, trochanter, mid leg trochanter, hind leg tarsus, anal cerci.

Yellowish Brown

Labrum, maxillary palpi, hind wing, hind leg tibia.

Holotype

Female, India, Maharashtra, Coll. 16-IX-2007; Awate Manisha; antenna, wings, cerci, ovipositor on slide labeled as above; other parts with pinned form, labeled as above.

Paratype

4♂, 2♀, Coll. From 2-VII-2006 to 2-XII-2007, same data as above.

Distributional Record

1♂ Shahuwadi, 6-VII-2006 1♂ Gaganbawada, 15-III-2006; 1♂ Gadhinglaj, 4-XI-2006; 1♂ Shirol, 17-VI-2007; ♀ Radhanagri, 3-VI–2007; 1♀ Kolhapur, 15-VIII-2007.

Etymology

As per given under description of male.

Remarks

According to the key of Chopard (1969) and Vasanth (1993) *Gryllus indica* sp. nov. runs close to *Gryllus bimaculatus* De Geer by having following characters.

1. Body glaborous black colour.
2. Yellow spot at base of tegmina.
3. Wings much longer than tegmina.

However, it differs from the above species by having following characters.

1. Ovipositor shorter than posterior femora.
2. Body length considerably smaller than *G. bimaculatus.*
3. Proportion of body length and antenna.

4. Tibia yellowish brown with black spots.
5. Flagellar formula
   1 = L/W = 0.61, 15 = L/W = 0.69, Ts = L/W = 2.30, A = 1.20

Gryllus pentagoni sp.nov.

Male (Figure 86)

Body elongated, yellowish to dark brown, very convex, 18.00 mm long excluding anal appendages, head 4.00 mm long, thorax 6.00 mm long, tegmina 11.00 mm long, hind wing 18.00 mm long; abdomen 8.00 mm long; anal cerci 6.00 mm long, dark brown, densely hairy.

Head

4.00 mm long, 5.00 mm broad, oval; occiput dark brown to blackish; compound eyes faint brown, ocular distance 0.95 mm, interocular distance 3.00 mm; 3 ocelli in triangle, yellowish brown, circular, no yellowish spot between the ocelli and eyes, ocellar distance 0.20 mm, interocellar distance 0.97 mm; head with wide frontal rostrum, frontal rostrum never with prolongation, width of frontal rostrum always about double that of first antennal segment, antennae on frontal rostrum, face hairy; clypeus black, triangular, 0.84 mm long, 1.27 mm broad; labrum dark brown 1.20 mm long, 2.00 mm broad; maxillary palpi yellowish to dark brown, cylindrical, 4.00 mm long, 0.28 mm broad, 5 segmented.

Antenna (Figure 87)

19.00 mm long, 0.76 mm broad, reddish brown; scape 0.47 mm long, 0.52 mm broad; pedicel 0.19 mm long, 0.20 mm broad; flagellum 152 segmented, 18.34 mm long, 0.13 mm broad; 1st segment 0.12 mm long, 0.13 mm broad; 15th segment 0.09 mm long, 0.14 mm broad; terminal segment 0.09 mm long, 0.05 mm broad.

Flagellar Formula

1 = L/W = 0.92, 15 = L/W = 0.64, Ts = L/W = 1.8, A = 1.12

Thorax

6.00 mm long, 5.00 mm broad, yellowish to dark brown, densely hairy, pronotum convex.

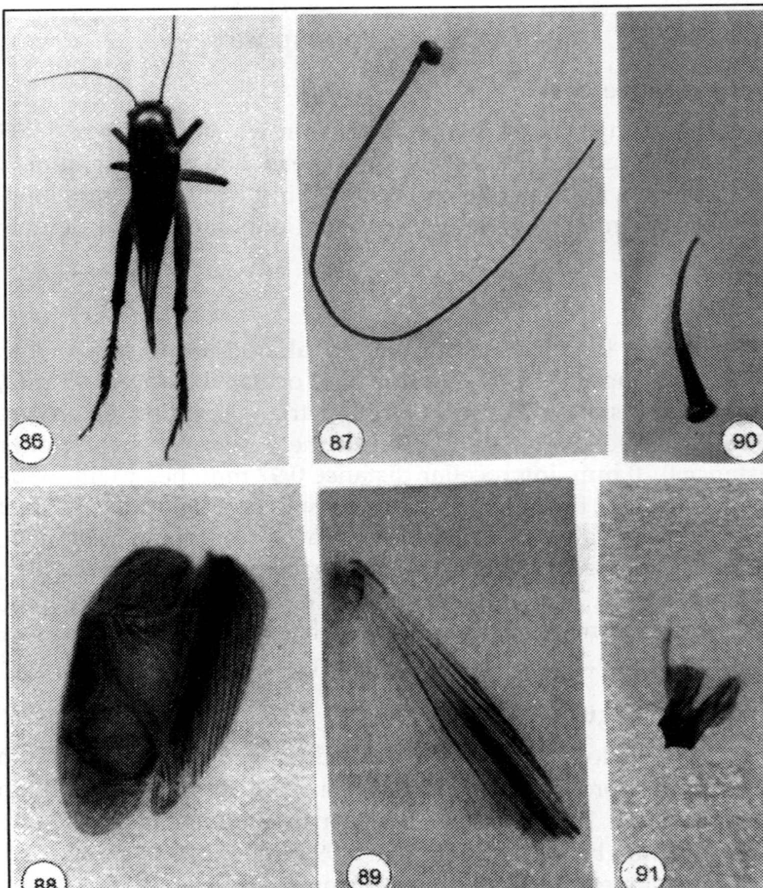

Plate 14: *Gryllus pentagoni* sp.nov.
Figure 86: Adult: Male; Figure 87: Antenna; Figure 88: Tegmina; Figure 89: Hind Wing; Figure 90: Anal Cerci; Figure 91: Genitalia

Tegmina (Figure 88)

Tegmina without whitish spot, long, narrow, extended to the apex of the abdomen, 11.00 mm long, 6.00 mm broad, yellowish brown, tegmina of male with well developed mirror, short, wide, mirror little longer than broad, nearly pentagonal and not rounded, apical field large presenting 6 veins, subcosta bearing 8 branches.

Hind Wing (Figure 89)

18.00 mm long, 7.00 mm broad, yellowish brown, wings long, longer than tegmina.

Legs

Fore leg, mid leg lighter than body colour, hind leg darker than body colour.

Fore Leg

11.01 mm long, 1.20 mm broad; coxa 1.32 mm long, 1.20 mm broad, dark brown, densely hairy at outer margin; trochanter 0.69 mm long, 0.52 mm broad, dark brown, sparsely hairy; femur 3.00 mm long, 1.00 mm broad, dark brown, densely hairy; tibia 3.00 mm long, 0.46 mm broad, yellowish to dark brown, 3 dark brown tibial spines, densely hairy; tarsus 3 segmented, 3.00 mm long, 0.29 mm broad, dark brown, densely hairy, dark brown tarsal denticles in a row; 1st tarsal segment 1.35 mm long, 0.60 mm broad; 2nd tarsal segment 0.45 mm long, 0.29 mm broad; 3rd tarsal segment 0.76 mm long, 0.21 mm broad; claw dark brown.

Mid Leg

12.38 mm long, 1.09 mm broad; coxa 0.86 mm long, 1.09 mm broad, dark brown, densely hairy; trochanter 0.52 mm long, 0.66 mm broad, yellowish brown, densely hairy; femur 4.00 mm long, 1.00 mm broad, dark brown, densely hairy; tibia 4.00 mm long, 0.45 mm broad, dark brown, densely hairy, 3 dark brown tibial spines; tarsus 3 segmented, 3.00 mm long, 0.34 mm broad, dark brown, densely hairy, reddish brown tarsal denticles in a row; 1st tarsal segment 1.04 mm long, 0.20 mm broad; 2nd tarsal segment 0.39 mm long, 0.28 mm broad; 3rd tarsal segment 1.20 mm long, 0.34 mm broad; claw yellowish brown.

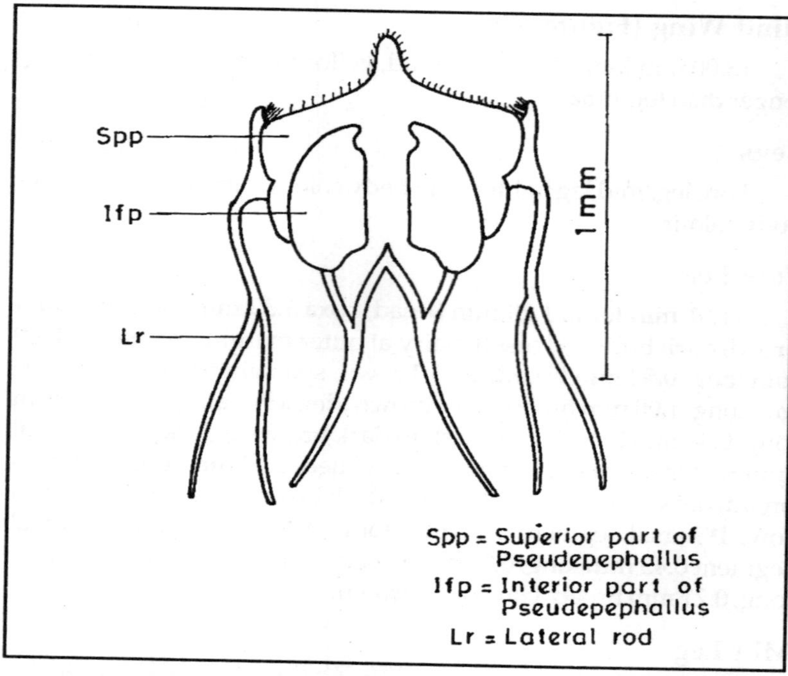

Figure 92: Genitalia of *Gryllus pentagoni* sp.nov.

Hind Leg

Posterior leg remarkably long, 22.39 mm long, 1.00 mm broad; coxa 2.00 mm long, 1.00 mm broad, yellowish brown, densely hairy at outer margin; trochanter 1.39 mm long, 1.35 mm broad, yellowish brown, sparsely hairy; posterior femora relatively slender, femur 11.00 mm long, 3.00 mm broad, yellowish to dark brown, densely hairy; tibia 9.00 mm long, 0.45 mm broad, yellowish to dark brown, densely hairy, armed with 16 reddish brown tibial spines; tarsus 3 segmented, 6.00 mm long, 0.64 mm broad, dark brown, yellowish brown tarsal denticles in a row; 1st tarsal segment 4.00 mm long, 0.30 mm broad; 2nd tarsal segment 0.41 mm long, 0.33 mm broad; 3rd tarsal segment 1.33 mm long, 0.20 mm broad; claw dark brown.

Abdomen

Abdomen 8.00 mm long, 4.00 mm broad; tapering to posterior side, 10 segmented, brown above yellowish beneath; anal cerci (Figure 90) 6.00 mm long, dark brown, densely hairy.

Genitalia (Figure 91 and 92)

Superior part of pseudepiphallus 1.16 mm long, 0.28 mm broad; lateral rod 1.33 mm long, 1.13 mm broad, lateral rod less acute; inferior part of pseudepiphallus 0.38 mm long, 0.1 mm broad, pointed, reddish brown.

Colour

Dark Brown to Blackish

Occiput.

Black

Clypeus.

Faint Brown

Compound eyes.

Yellowish Brown

Ocelli, tegmina, hind wing, mid leg trochanter, claw, hind leg coxa, trochanter, tarsal denticles.

Reddish Brown

Antennae, mid leg tarsal denticles, hind leg tibial spines.

Yellowish to Dark Brown

Maxillary palpi, thorax, fore leg tibia, hind leg femur, tibia, abdomen.

Dark Brown

Labrum, fore leg coxa, trochanter, femur, tibial spines, tarsus, claw, mid leg coxa, femur, tibia, tibial spines, tarsus hind leg tarsus, claw, anal cerci.

Holotype

Male, India, Maharashtra, Coll. 27-X-2006; Awate Manisha; antenna, wings, cerci, genitalia, on slide labeled as above; other parts with pinned form, labeled as above.

Paratype

8 ♂, 3 ♀, Coll. from 3-VI-2006 to 30-XI-2007, same data as above.

Distributional Record

3♂ Gaganbawada, 15-VIII-2006; 2♂ Kolhapur, 10-X-2006; 1 Kagal 29-VI-2006; 2♂ Hatkangale, 31-VIII-2006; 1♂ Ajara, 20-VIII-2007; 2♀ Bhudargad, 15-VI-2007.

Etymology

The mirror of the species is more or less pentagonal in shape hence the name *Gryllus pentagoni* sp. nov.

Remarks

According to the key of Chopard (1969) and Vasanth (1993) *Gryllus pentagoni* sp. nov. runs close to *Gryllus gracilipes* Saussure by having following characters.

1. Abdomen brown above yellowish beneath.
2. Tegmina without whitish spot.
3. Femora slender.

However, it differs from the above species by having following characters.

1. No yellowish spot between the ocelli and eyes.
2. Posterior tibiae shorter than femora.
3. Posterior tibia armed with 8 spines instead of 6 to 7 spines.

4. Mirror is more or less pentagonal not rounded.
5. Apical field presenting 6 veins.
6. Subcosta bearing 8 branches instead of 5.
7. Inferior part of pseudepiphallus pointed and lateral rod less acute.
8. Flagellar formula

   $1 = L/W = 0.92, 15 = L/W = 0.64, Ts = L/W = 1.8, A = 1.12$

Gryllus lanceolati sp.nov.

Female (Figure 93)

Body colouration black, convex, elongate, pubescent, 15.00 mm long excluding anal appendages, head 3.00 mm long, thorax 5.00 mm long, tegmina 8.00 mm long, hind wing 15.00 mm long; abdomen 7.00 mm long, blackish; anal cerci 7.00 mm long, yellowish brown; ovipositor 11.00 mm long, reddish brown.

Head

Head shining with a narrow yellow band, 3.00 mm long, 4.00 mm broad, rounded; occiput brownish black, head adorned with light bands of the occiput; compound eyes faint brown, ocular distance 1.21 mm, interocular distance 2.00 mm; head with tansverse yellow band between ocelli, 3 ocelli disposed as a triangle, yellowish white, circular, ocellar distance 0.34 mm, interocellar distance 0.86 mm; head with wide frontal rostrum, antennae on frontal rostrum, face convex; clypeus dark brown to blackish, triangular, 1.08 mm long, 0.62 mm broad; labrum yellowish brown, semicircular, 1.00 mm long, 2.00 mm broad; maxillary palpi yellowish to dark brown, cylindrical, 4.00 mm long, 0.28 mm broad, 5 segmented.

Antenna (Figure 94)

16.00 mm long, 0.57 mm broad, yellowish brown; scape 0.52 mm long, 0.57 mm broad; pedicel 0.22 mm long, 0.13 mm broad; flagellum 157 segmented, 17.00 mm long, 0.10 mm broad; 1st segment 0.10 mm long, 0.17 mm broad; 15th segment 0.10 mm long, 0.12 mm broad; terminal segment 0.08 mm long, 0.05 mm broad.

Flagellar Formula

$1 = L/W = 0.58, 15 = L/W = 0.83, Ts = L/W = 1.6, A = 1.00$

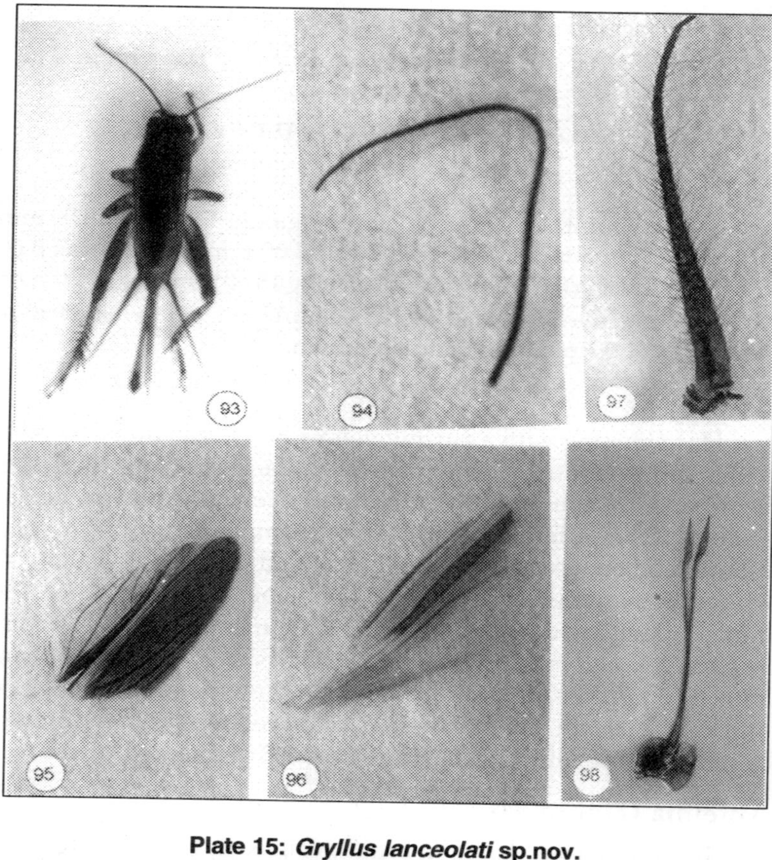

Plate 15: *Gryllus lanceolati* sp.nov.
Figure 93: Adult: Female; Figure 94: Antenna; Figure 95: Tegmina; Figure 96: Hind Wing; Figure 97: Anal Cerci; Figure 98: Ovipositor

Thorax

6.00 mm long, 4.00 mm broad, dark brown to black, densely hairy, pronotum dark, lateral lobes rectangular.

Tegmina (Figure 95)

Female with more perfectly developed tegmina, never reduced to small lateral pads, tegmina long, narrow, not covering the tip of abdomen, 8.00 mm long, 4.00 mm broad, blackish.

Hind Wing (Figure 96)

15.00 mm long, 5.00 mm broad, yellowish brown, wings longer than tegmina.

Legs

Fore leg lighter than body colour; mid leg, hind leg darker than body colour.

Fore Leg

7.65 mm long, 0.95 mm broad; coxa 1.26 mm long, 0.95 mm broad, yellowish brown, densely hairy at outer margin; trochanter 0.74 mm long, 0.69 mm broad, yellowish brown, sparsely hairy at outer margin; anterior femora with yellowish marking above, femur 2.00 mm long, 1.00 mm broad, yellowish brown, densely hairy; tibia 2.00 mm long, 0.41 mm broad, 5 yellowish brown tibial spines, yellowish brown, densely hairy; tarsus 3 segmented, 1.65 mm long, 0.26 mm broad, yellowish brown, densely hairy, denticles present in rows; 1st tarsal segment 0.83 mm long, 0.20 mm broad; 2nd tarsal segment 0.37 mm long, 0.23 mm broad; 3rd tarsal segment 0.26 mm long, 0.20 mm broad; claw yellowish brown.

Mid Leg

11.55 mm long, 0.74 mm broad; coxa 0.86 mm long, 0.74 mm broad, dark brown, densely hairy at outer margin; trochanter 0.69 mm long, 0.66 mm broad, yellowish brown, sparsely hairy; femur 3.00 mm long, 0.74 mm broad, dark brown to blackish, densely hairy; tibia 4.00 mm long, 0.34 mm broad, dark brown, densely hairy,dark brown 5 tibial spines; tarsus 3 segmented, 3.00 mm long, 0.26 mm broad, yellowish to dark brown, densely hairy, denticles present in rows; 1st tarsal segment 2.00 mm long, 0.3 mm broad; 2nd tarsal segment 1.16 mm long, 0.18 mm broad; 3rd tarsal segment 0.32 mm long, 0.21 mm broad; claw reddish brown.

Hind Leg

19.52 mm long, 3.00 mm broad; coxa 2.00 mm long, 3.00 mm broad, yellowish brown, sparsely hairy; trochanter 0.52 mm long, 1.30 mm broad, dark brown, sparsely hairy; femur 7.00 mm long, 2.08 mm broad, yellowish brown to dark brown, densely hairy, posterior femora no longer than tibia and metatarsus together, blackish oblique streaks on the hind femora; posterior tibiae non serrulated at base, tibia 6.00 mm long, 0.48 mm broad, dark brown, 10 dark brown tibial spines, densely hairy; tarsus 3 segmented, 4.00 mm long, 0.29 mm broad, dark brown, densely hairy, denticles present in rows; 1st tarsal segment 2.00 mm long, 0.39 mm broad; 2nd tarsal segment 1.25 mm long, 0.17 mm broad; 3rd tarsal segment 0.41 mm long, 0.22 mm broad; claw reddish brown.

Abdomen

7.00 mm long, 3.00 mm broad, tapering to posterior side, dark brown to blackish, 9 segmented; anal cerci (Figure 97) 7.00 mm long, yellowish brown; ovipositor (Figures 98 and 99) 11.00 mm long, 0.43 mm broad, considerably shorter than body, reddish brown, lanceolate.

Colour

Brownish Black

Occiput.

Faint Brown

Compound eyes.

Yellowish White

Ocelli.

Blackish

Tegmina

Yellowish Brown

Labrum, antennae, hind wing, fore leg coxa, trochanter, femur, tibia, tibial spines, tarsus, claw, mid leg trochanter, anal cerci, hind

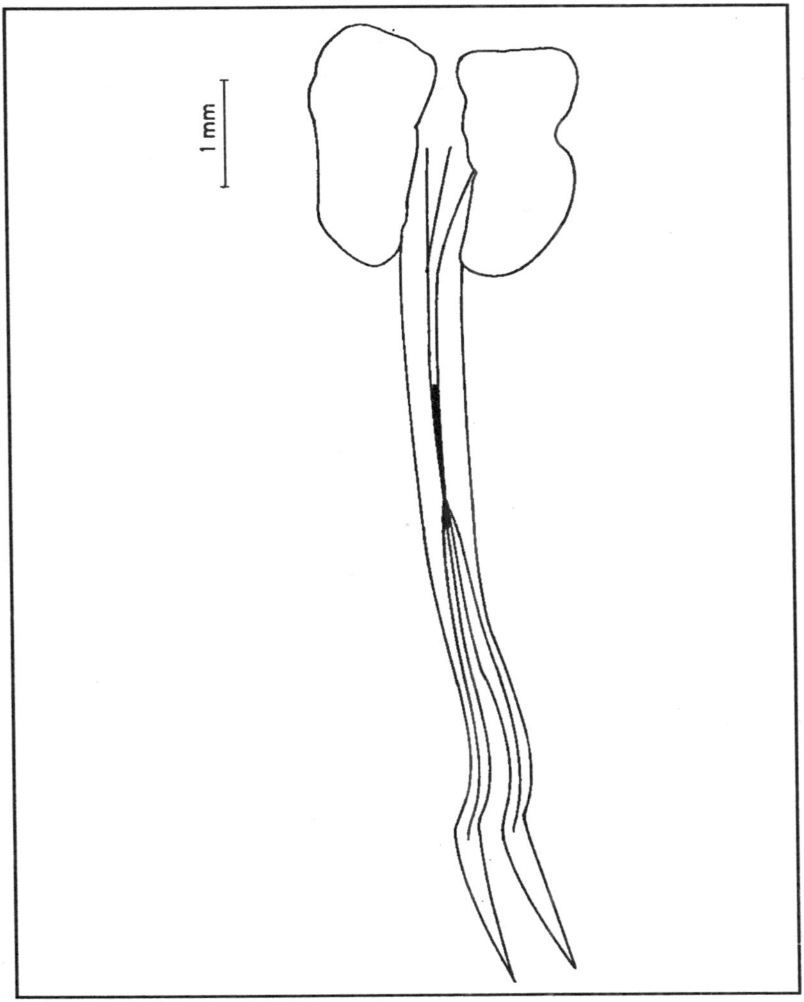

Figure 99: Ovipositor of *Gryllus lanceolati* sp.nov.

leg coxa.

Dark Brown to Blackish

Clypeus, thorax, mid leg femur, abdomen.

Dark Brown

Mid leg coxa, tibia, hind leg trochanter, tibia, tibial spines, tarsus.

Yellowish to Dark Brown

Maxillary palpi, mid leg tarsus, hind leg femur.

Reddish Brown

Mid leg claw, hind leg claw, ovipositor.

Holotype

Female, India, Maharashtra, Coll. 17-IX-2007; Awate Manisha; antenna, wings, cerci, ovipositor on slide labeled as above; other parts with pinned form, labeled as above.

Paratype

7♂, 2♀, Coll. from 3-VI-2006 to 30-XI-2007, same data as above.

Distributional Record

2♂ Shirol, 6-VII-2006; 3♂ Panhala, 23-VII-2006; 1♀ Panhala 23-VII-2006; 2♂ Kagal, 16-IX-2007; 1♀ Kolhapur, 26-VIII-2007.

Etymology

The ovipositor of the species is lanceolate hence the name *Gryllus lanceolati* sp. nov.

Remarks

According to the key of Chopard (1969) and Vasanth (1993) *Gryllus lanceolati* sp. nov. runs close to *Gryllus signipes* Walker by having following characters.

1. Abdomen brown above yellowish beneath.
2. Tegmina without whitish spot.
3. Femora slender.

However, it differs from the above species by having following characters.

1. Black body colouration.
2. Blackish streaks on the hind femora.
3. Head shining with a narrow yellow band.
4. Pronotum dark.
5. Tegmina is blackish no brown and not covering the tip of abdomen.
6. Ovipositor considerably shorter than body.
7. Proportion of cerci to ovipositor.
8. Proportion of posterior tibia to femur.
9. Anterior femora with yellowish marking above instead of black.
10. Flagellar formula

    $1 = L/W = 0.58, 15 = L/W = 0.83, Ts = L/W = 1.6, A = 1.00$

Chapter 2
Household Pests

1. Cockroaches

Dictyoptera
Blattidae
***Periplanata americana* Linnaeus**

(*i*) *P. americana*

Distribution

The cockroach, *P. americana* is world wide in distribution. It is widely distributed in India, almost in every state of the country.

Marks of Identification

P. americana (Figure 100) is largest species found in India. It is dark brown in colour and measures about 25 mm in body length and 11 mm in width. It is abundant in southern parts of our country including Maharashtra. Eggs of this species are brownish barrel shaped and found attached to any substrate where they stay for longer period. Nymphs are flat bodied, brownish, without wings, similar to adults by shape but they are miniatures. Pronotal plate have 4 spots towards posterior side. A pair of cerci is present on the last abdominal segment. Nymphs are crawlers.

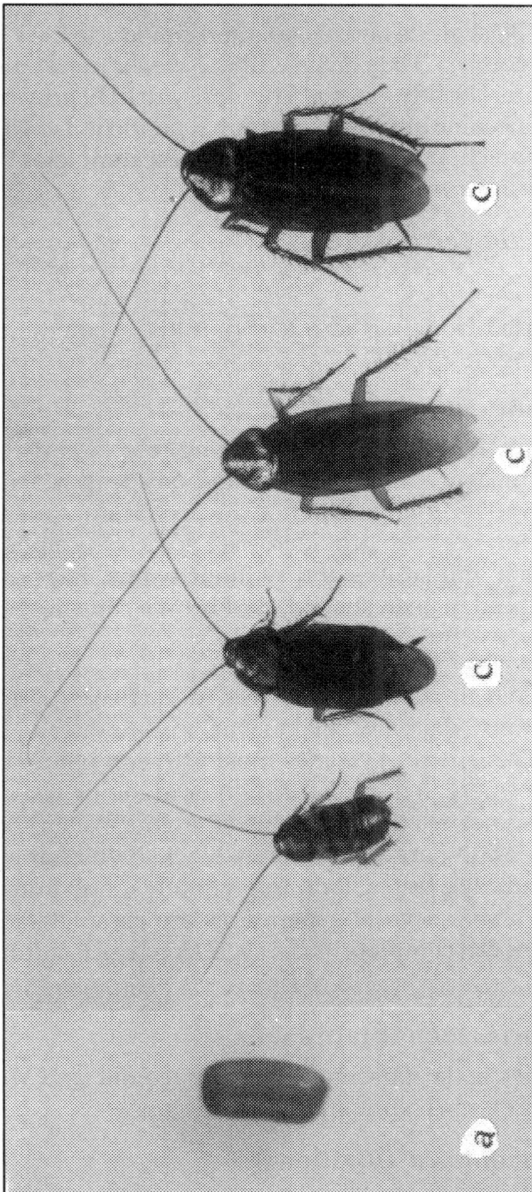

Figure 100: Cockroach (*Periplanata americana*); Life Cycle (a) Egg (Oothica); (b) Nymph; (c) Adult

Life Cycle (Figures 100a,b,c)

About 20-30 eggs are laid in craks and crevices in the kitchen in purse shaped ootheca. They hatch within 3 weeks in summer and 3 months in winter. Newly emerged nymphs are light gray coloured but, soon they become brownish. Nymphal period is 3 to 10 months depending upon climatic conditions. The pest completes only one generation during a year.

(ii) *Shelfordella tartara* Adel.

Distribution

It is reported from northern part of India as common pest in human dwelling.

Marks of Identification

S. tartara in its adult stage is shining, black, and flattened individual. It measures about 29 mm in body length. Its head is bowed downward and covered totally by pronotum. Antennae are long and slender. Legs are provided with spines. Nymphs are flat bodied, blackish, miniatures and without wings. Eggs are cigar shaped and placed in bean shaped, darkbrown capsule.

Life Cycle

Eggs are laid in craks and crevices in kitchen or latrine, etc. They hatch within 2 days and several weeks in summer and winter respectively. This species becomes most active in hot and humid season. Newly emerged nymphs are lighter in colour and cared by the mother for some period. Approaching the older instars, the colour becomes darker and darker and finally black. Nymphal period is about one month. The last instar numph finally moult into winged adult. During a year several generations are completed. During summer the pest can complete its life cycle from egg to adult within 4 to 5 weeks.

(iii) *Blatta orientalis* (Blattidae)

This species is dark brown to black in colour and wingless. Nymphs are lighter in colour and smaller in size.

(iv) *Blatta germanica* (Blattidae)

This species is small sized. It shows pale ocherous body colouration. Its pronotum shows two longitudinal lines. Nymphs also shows two longitudinal dark broad lines on pronotum and on few abdominal segments and matching with large black patch of abdomen.

(*v*) *Supella supellectium* (Blattidae)

It is small sized species with brown and white bands on the body.

Nature of Damage

Nymphs as well as adults of cockroaches eat all kinds of food and produce excreta on food material and elsewhere in the kitchen or house. Thus, contaminate our food. Cockroaches also feed on starchy content of book binding and thus spoil the books. Now a days, cockroaches are also breeding in latrines and in craks and crevices of bath room in hostels, hotels and guest houses. They come out at night and crawl all over the unprotected foods for feeding purpose and contaminate with their excreta and latrine filth. They also leave unacceptable roachy odour in houses. It is suspected that they have disease transmitting role by contaminating food with pathogens. Cockroaches are thus, nuisance in trains, restaurants, guest houses, kitchens and old musty buildings.

Control Measures

1. The pest can be controlled by spraying haunts and target places with malathion 0.2 per cent or dusting 5 per cent malathion or carbaryl.
2. The pest can be controlled by keeping the objects clean since cockroaches like dark moist places, boxes, cupboards, crevices, drains, etc.
3. Baiting with 2 per cent yellow or white phosphorus paste or granules.

Natural Control

1. Evanid fly, *Evania* sp. (Evanidae) parasitizes ootheca of cockroach and cause mortalities.
2. *Ampulex* sp. parasotizes adult cockroaches and kill by developing upon them.

2. Silverfish (Figure 101) **Thysanura**

Lepismidae

***Lepisma saccharina* Linnaeus**

(*i*) *L. saccharina*

Distribution

There are about 550 species of silverfishes in the world. From India, 25 species have been reported (Sathe, 2003). *L. saccharina* is cosmopolitan in distribution. From India it is widely reported from all states.

Marks of Identification

L. saccharina is uniformly silvery white and carrot shaped. It is soft bodied wingless insect measuring about 13 mm in body length. Silver scales are scattered all over the body of the insect. It shows long antennae and caudal filaments. Head is 2 mm long, prognathus,

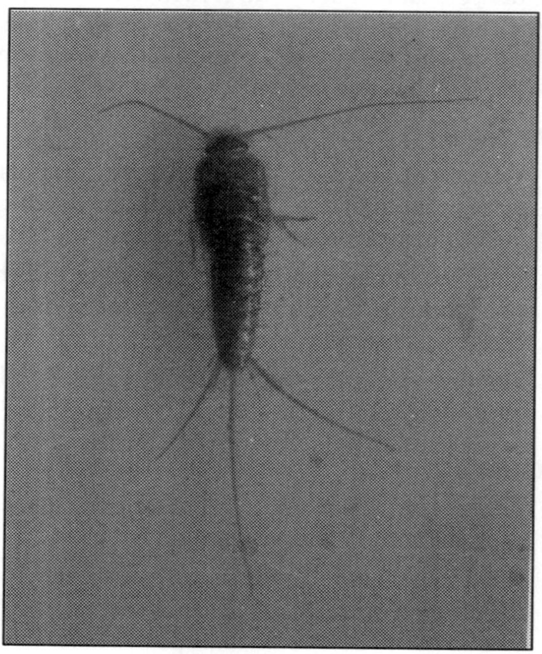

Figure 101: Silverfish

roughly triangular, dark yellow and hairy at the tip. Antennae are 8.25 mm long and 0.26 mm wide. Thorax is 4.00 mm long, yellowish, hairy laterally, 2.5 mm broad. Fore leg is 6.98 mm long, yellowish, hairy and shorter than body. Coxa is 2.2 mm long, trochanter 0.44 mm long, femur 1.6 mm long, tibia 1.3 mm long, tarsus 1.3 mm, metatarsus 0.5 mm long and claw is with two spines. Abdomen is 6.0 mm long and yellwish white. Caudal filament is 8.00 mm long and more than 50 segmented. Anal cerci are 8.60 mm long. Silvery scales are present all over body.

(*ii*) *Termobia domestica* Packard

Distribution

It is distributed in Maharashtra, Karanataka, Andra Pradesh, Madhya Pradesh, etc.

Marks of Identification

Adult silverfish is 15 mm long. Its head is black and thorax and abdomen is with blackish and whitish bands. Head is 1.50 mm long and prognathus. Eyes are black. Antenna is 13.2 mm long, dark black with pale white bands. Fore leg is 7.75 mm long, pale white with black bands. Caudal filaments is 12.75 mm long and anal cerci is 11.00 mm long.

Life Cycle

Mated female lays few eggs in cracks, crevices and warm and damp places. They hatch within 6-8 days. Newly hatched nymphs resembles with adults by shape but are very small in size. The nymph moults for indefinite times. Thus, its development is quite slow. The nymphal period is 7-9 months. Over lapping generations are common in silverfishes.

Nature of Damage

Silverfishes are associated with old calendars, old books, files, papers neglected or dumped for longer period. They feed on papers and book bindings and damage old and rare books. They also feed on clothes.

Control Measures

1. Dusting the pest habitats with 5 per cent BHC or Malathion.

2. Using poisoned glue madeup as a mixture containing wheat flour 200 pts, Sodium fluoride 16 pts, Powdered sugar 10 pts, Powdered common salt 5 pts and water to make paste.
3. Cleaning libraries and dumping material periodically will reduce pest population.

3. Booklouse

Pscocidae
Psocoptera
***Liposcellis transvalensis* Ender.**

Distribution

The booklouse is cosmopolitan in distribution. It is widely distributed in India, Pakistan, Bangladesh, China, etc.

Marks of Identification

The booklouse is wingless, soft bodied, tiny insect found in old books, behind old calendars on wall and wall papers, etc.

It is active insect of light brown or pale yellow colour which shows chewing type of mouth parts. The book louse has longer legs and antennae and louse like shape. Nymphs are similar to adults in shape but, are smaller in size. The adult measures about 1 mm in body length.

Life Cycle

Mated female lays good number of eggs either singly or in groups at damp places near to books or on books in libraries. The eggs hatch within 4 to 6 days in favorable conditions. The newly emerged nymphs are gregarious and found feeding on book bindings. The nymph moults for several times. The number of nymphal instars are dependant on climatic conditions. If worm conditions available, the pest breed throughout the year. Overlapping generations are common in the pest.

Nature of Damage

Both, nymphs and adults feed on starchy (glued) content of book bindings and destroy old books, old-calendars, old paper pictures on wall and old photographs by feeding upon them. The booklice can also damage botanical and zoological specimens in museums. Flour and cereal products are also damaged by this insect by feeding upon them in store houses or in kitchen rooms.

Control Measures

Use poison baits prepared as follows:

1. Sodium fluoride plus flour 1:8 (or)
2. Use fumigants like So2 (or)
3. The room/library should be heated to 50-60 °C.
4. Almirahs cupboards should be kept clean.

4. Furniture Beetles (Figure 102c)

Coleoptera
Bostrichidae
Heterobostrychus aequalis

(*i*) *Heterobostrychus aequalis*

Distribution

H. aequalis is widely distributed in India.

Marks of Identification

The adult is small beetle which is cylindrical and black in colour. Its prothorax is typically extending over head as a hood and hind margin of elytra bears curved projection. The grub is whitish yellow with dark head and strong mandibles. Pupa is dark brown and exarate type.

Life Cycle

The adult beetle lay eggs in egg chamber or tunnel prepared by female in timber or on bark. The eggs hatch within a week or so. Newly emerged grubs bore into the wood. The grub prepare irregular tunnellings into the wood which are filled with fine wood dust. At the end of tunnel, the full grown grub pupates. When pupa is transformed into adult, the adult also feed on wood content by making tunnel and finally come out by making an exit on the surface of wood. Larval life is 6 months and pupal life is 6 months. Thus, the life cycle is completed with 1 to 6 years depending upon climatic conditions.

Nature of Damage

The beetle and grub cause damage to furniture or wooden structures in the houses.

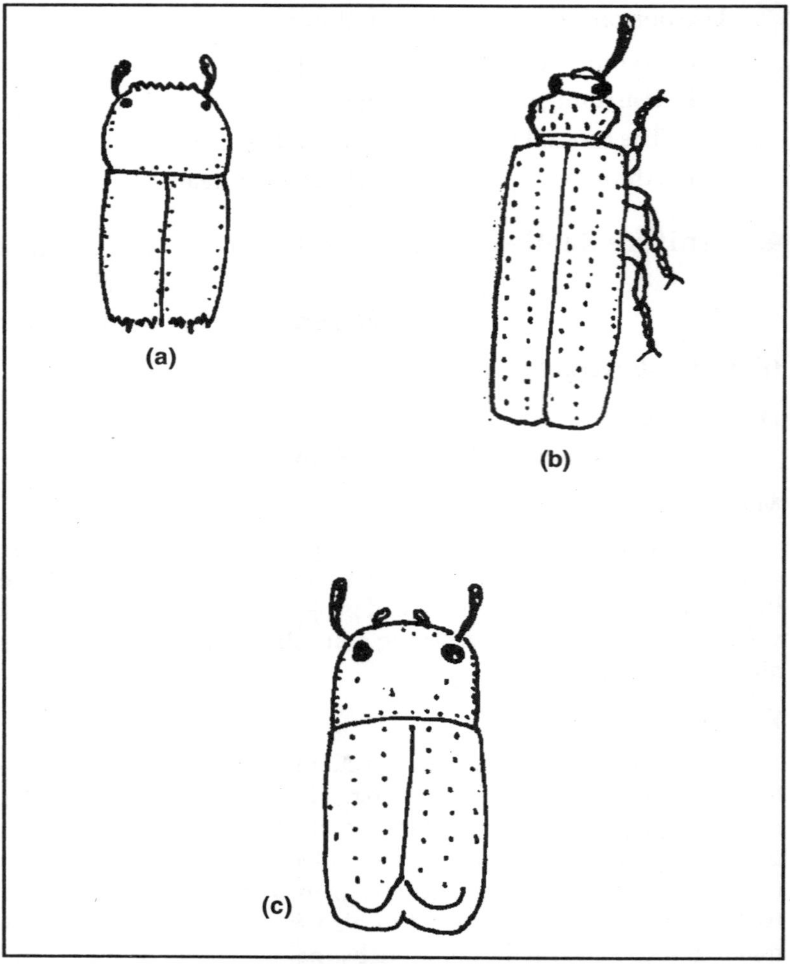

Figure 102: Furniture Beetles
(a) *Dinoderus ocellaris*; (b) *Lyctus africanus*;
(c) *Heterobostrychus aequalis*

(*ii*) *Lyctus africanus* Lesne (Figures 102b) Coleoptera
Bostrichidae
***Lyctus africanus* Lesne**

Distribution

Asia, Africa, etc. In India, it is widely distributed in various states. This beetle is known as "powder-post" beetle.

Marks of Identification

L. africanus is light brown beetle measuring about 3-4 mm in body length. Its prothorax is broader at anterior side and narrower towards posterior, scutellum is shining; small hairs present on elytra. Grub is yellowish white with strong mandibles. Pupa is brownish and exarate type.

Life Cycle

Mated female lays her eggs in open holes of the sawn timber or furniture. Eggs hatch within 1 to 2 weeks. Newly emerged grubs bore into the wood. Later, the grubs borer deeper and widen the tunnel. The tunnel is packed with powdery frass excreted by the grubs. Full grown grub pupate at the end of tunnel. Then pupa is moult into adult in the same tunnel. The adult thus, formed continue tunneling for feeding and for coming out of the wood or furniture. In adverse condition the life cycle take 8 years time but, in favorable condition, the pest completes 3 generation in a year.

Nature of Damage

Grubs as well as adults tunnel the wood. The furnitures and drywood affected by this pests shows tunneling filled with powdery frass. The pest attacks telephonic or other wooden poles. Packing cases, planks, board furniture plywood, tool handles, etc are also attacked by this pest species.

(*iii*) *Dinoderus ocellaris* Stephens (Figure 102a) Coleoptera
Bostrichidae
***Dinoderus ocellaris* Stephens**

Distribution

This species is distributed all over India.

Marks of Identification

The beetle is brownish which measures about 3-4 mm long in body length. Antenna is 10 segmented and clubbed. Front clypeal region is fringed with hairs. Grub is yellowish white and curved. Mandibles are black coloured. Abdominal segments are yellowish.

Life Cycle

Mated female lays good number of eggs in pores and depressions on furniture surface or wood surface. Eggs hatch within 5-6 days. Newly emerged grubs bore into the wood of bamboo. The tunnels are filled by powdered excreta and frass. The full grown grub pupate at the end of tunnel. The pupal period is 3-6 months. Newly formed adult stay in the tunneling for feeding and then coming out of the wood or furniture by making a circular hole. The adult also travel back with the same hole and tunnel in other direction freshly in the wood or furniture. The pest completes its life cycle within 1 to 2 years.

Nature of Damage

Both, adults and grubs damage bamboo furniture and bamboo articles. The pest also damage felled trees of bamboo in forest ecosystem and in plains in human dwellings. The estimated damage to furniture and drywood costs is several lakhs in every season and year.

Control Measures

1. Furniture should be painted with benzene or CCl_4 mixed with naphthalene.
2. Furniture should be polished properly with damage preventives.
3. Furniture should be polished or varnished periodically for prevention of pest attack.
4. Timber should deeped into $Cuso_4$ or $Zncl_2$ solution for get rid from *Lyctus* spp.
5. Infected wood or furniture be fumigated with CH_3Br in tent or closed vaqults.
6. Crude creosote oil is used for painting bamboo articles for protecting from powder post beetles. Furniture should be

polished or varnished periodically for prevention of pest attack.

7. Spraying wood or furniture with 2 per cent chlordane or 5 per cent DDT is also good pest management strategy against powder post beetles.

5. Carpet Beetles

Coleoptera
Dermestidae
Anthrenus scrophulariae

Distribution

India, Pakistan, Bangladesh, Indonesia, China, etc.

In India, it is widely distributed in several states.

Marks of Identification

The beetle measures about 3 mm in body length.

It is with yellow, red, white and black mottled markings. It has oval shape, retractile head and clubbed antennae. Tarsi are five segmented. The larva is elongate, cylindrical and tapering posteriorly and creamy white.

Life Cycle (Figure 103a–d)

Mated female lays her eggs on the wool or hairs or fur or leather or dry insects. Incubation period is 7 days. Newly emerged grubs start feeding on wool/hairs/fur/dry insects/leather specially at dark folds. The exuvae of larvae resembles with the living forms of grubs. The larva bore into feather/leather/dry insects/etc. The full grown grubs pupate in tunnel. The larval period is 18 months and pupal period is 24 days. The life cycle is completed within 1 to 3 years.

Nature of Damage

1. The larva feed on any animal product as wool, hairs, leather, skin, fur, feather, bristles, horns, silk rayon, dry insects, linen and jute, etc.
2. The larva found feeding in between folds of fabrics.
3. Larva make numerous holes of fabrics, above mentioned materials.

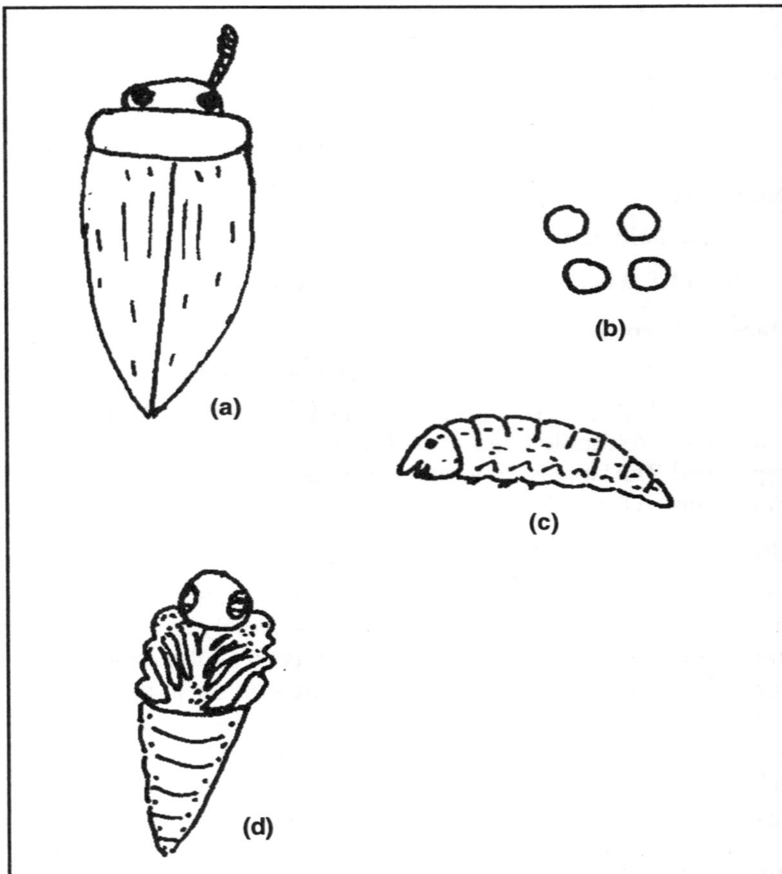

Figure 103: Carpet Beetles
(a) Adult; (b) Eggs; (c) Grub; (d) Pupa

4. Pest can damage museum specimens.
5. The larva also bore horns of animals.
6. The larva and adult damage insects in insect boxes.
7. The pest also damages plants.

Control measures

1. Articles to be protected should be exposed to strong sunlight.
2. Articles should be beaten and frequently brushed.
3. Use Naphthalene bolls.
4. Use polythene bags for keeping protective materials.
5. Specimen should be painted with turpentine oil containing 10-12 per cent Orthodichlrobenzene.
6. Bones and horns should be cleaned with terpentine oil.
7. Box insects and plant specimens should be controlled by a mixture of naphthalene powder, kerosene, chloroform and petroleum (6:1:1:4).
8. Use mixture of naphthalene powder and paradi-chlorobenzene.
9. In severe infestation, entire house be fumigated with CS_2.
10. Spray lindane/DDT into craks and crevices.
11. Pyrethrum mixed with kerosene be advised against this pest.
12. Use following bait against carpet beetles.

    A piece of old woolen cloth soaked in alcoholic extract of fish oil in a small box could be a good idea.

Attagenus piceus

Coleoptera
Desmestidae
Attagenus piceus

Distribution

It is found in India and in several other countries of the world.

Marks of Identification

The species is uniformly dull black which measures about 3-4 mm in body length.

Life Cycle and Control Measures

As above species.

6. Cigaratte Beetle

Coleoptera
Anabiidae
***Lasioderma serricorne* Fabr.**

Distribution

It is world wide in distribution.

Marks of Identification

The beetle is reddish brown in body colour and some what rounded, measuring about 2.6 mm in body length. Its head and thorax are bend downwardly and gives humpy look. Grubs are curved, hairy and yellowish with head brown colour. Pupa is brownish and exarate type. The pest is visualized by different common names as tobacco beetle or cigarette beetle or cheroot beetle.

Life Cycle (Figures 104a–d)

A single mated female can lay 100 eggs on dry tobacco leaves/ stored cigarettes. The eggs hatch within 6-10 days. Newly emerged tiny grubs start feeding on dried leaves of tobacco or cigarettes stored in store house. They pass through 6 to 7 instars. Larval period is 30-50 days depending on climatic conditions. The grubs bore into the stored tobacco material and make it powdered form. The full grown grub pupates in the tobacco product. For pupation grub prepares silken cocoon covered with food materials. Thus, pupation takes place in silken cocoon and lasts for 8-10 days. The pest completes 6 generations in a year. The pest becomes active from April to October.

Hosts

Spices crops such as tobacco (dry leaves or dry leaves products) and black and red pepper (seeds used in drugs).

Nature of Damage

Both grubs and beetles feed on tobacco dry leaves and tobacco products. The pest infests cigarettes in cigar factories if they stored for longer period. Attack start at the time of drying of leaves and harvesting of tobacco leaves. The pest also feed on seeds of spices

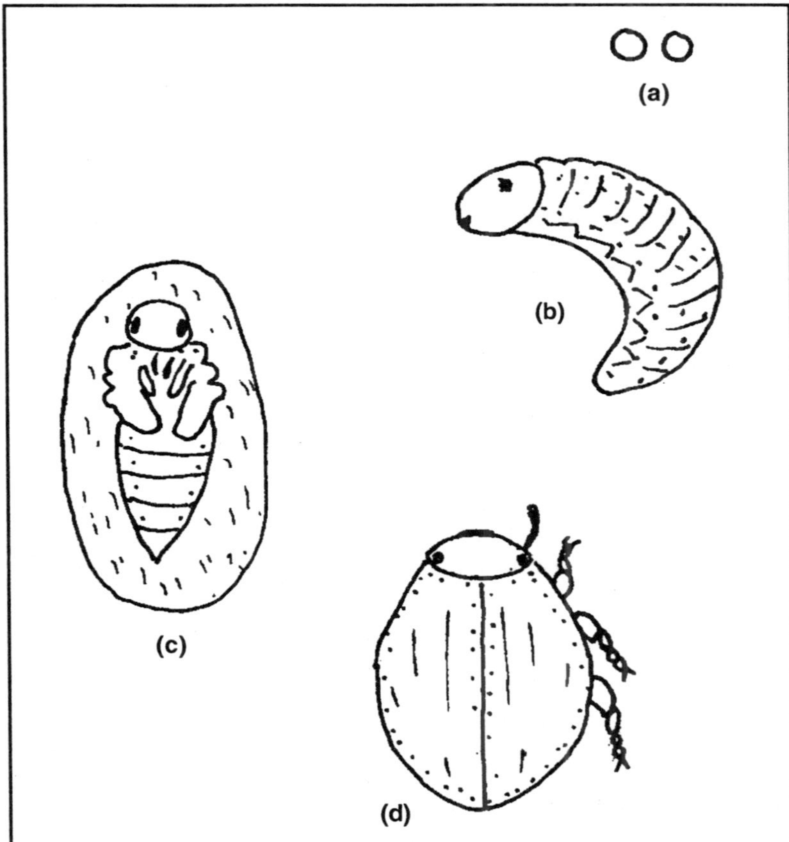

Figure 104: Cigaratte Beetles
(a) Eggs; (b) Grub; (c) Pupa in Cocoon; (d) Adult

plants such as black and red pepper. Seeds of these plants are used forming drugs.

7. Clothes Moths

Lepidoptera
***Tinaea* Tinaeidae**
(*i*) *Pellionella* L.
(*ii*) *Pachyspila* Meyr.

Distribution

It is found all over India.

Marks of Identification

Moths are straw coloured, small sized with wing expanse of 1.25 cm. Caterpillars are pale yellow. Pupae are brown and object type.

Life Cycle (Figures 105a–d)

Mated female lay her eggs on fabrics. Eggs hatch within 4-5 days. The female can lay about 75 eggs. Newly emerged tiny larvae feed on clothes. The larva prepares parchment like silken covering around itself. The larva moults for five times. Full grown larva pupate in larval covering only. Pupal period is about 10 days. The adults thus, formed mate soon and female start egg laying on various kinds of clothings.

Nature of Damage

They feed on clothings and woolen articles or woolen cloths. The larva also cause damage to hoofs, feathers, fur, dried skin and dried insects. The pest prepare numerous holes to above said goods.

Control Measures

1. Sun drying of infected materials.
2. Clothes should be packed in polythene bags.
3. Fumigation to boxes, houses, museums, stores, etc. should be done.
4. Spraying to infected parts with Eulan and Mittin FF.

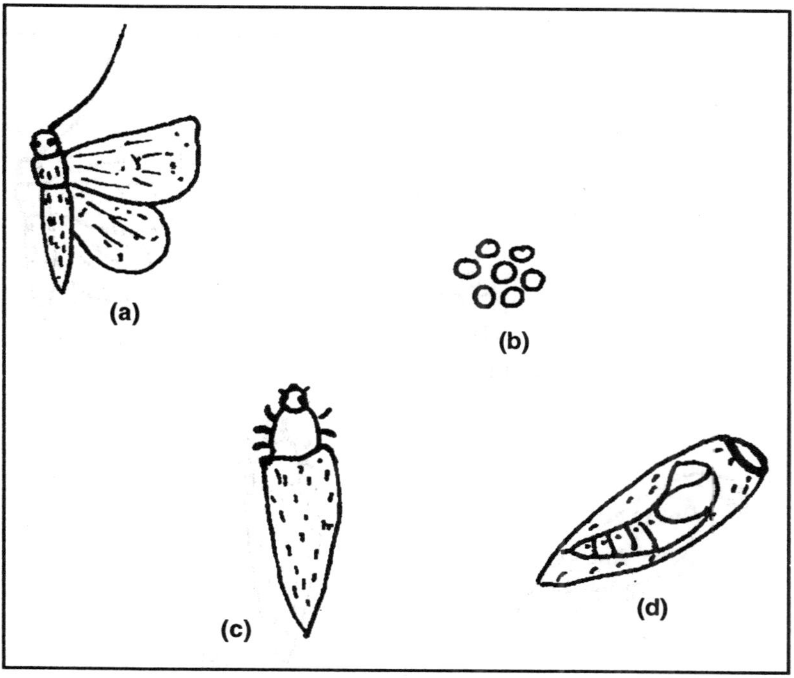

Figure 105: Clothes Moths–Life Cycle
(a) Adult; (b) Eggs; (c) Larva; (d) Pupa

8. Ants

Hymenoptera
Formicidae
***Monomorium indicum* Forel**

Distribution

This species is worldwide in distribution. In India, it is widely distributed in almost every state.

Marks of Identification

M. indicum adults are small sized, 2.5-3.00 mm in body length and brownish in body colouration. It has petiolate abdomen and wingless in appearance. Antennae are geniculate. Grubs are whitish, pupae are brownish and exarate type. It is social insect which shows caste system, polymorphism and division of labour. Thus, the ants are very disciplined insects.

(*i*) Black Ant

Hymenoptera
Formicidae
***Camponotus compressus* Fabr.**

Distribution

Cosmopolitan.

Marks of Identification

The species is large sized and black coloured. Its abdomen is petiolate and contain no wings. Grubs are whitish, without legs, pupae are blackish. Winged and non winged forms are common in this species.

(*ii*) *Monomorium indicum* Forel

Distribution

All over the world. In India it is reported from almost every state.

Marks of Identification

It is small sized and measuring about 3 to 3.5 mm in body length and brownish in colour. Morphologically grubs of this species are similar to *C. compressus* but are smaller in size. While, pupae are

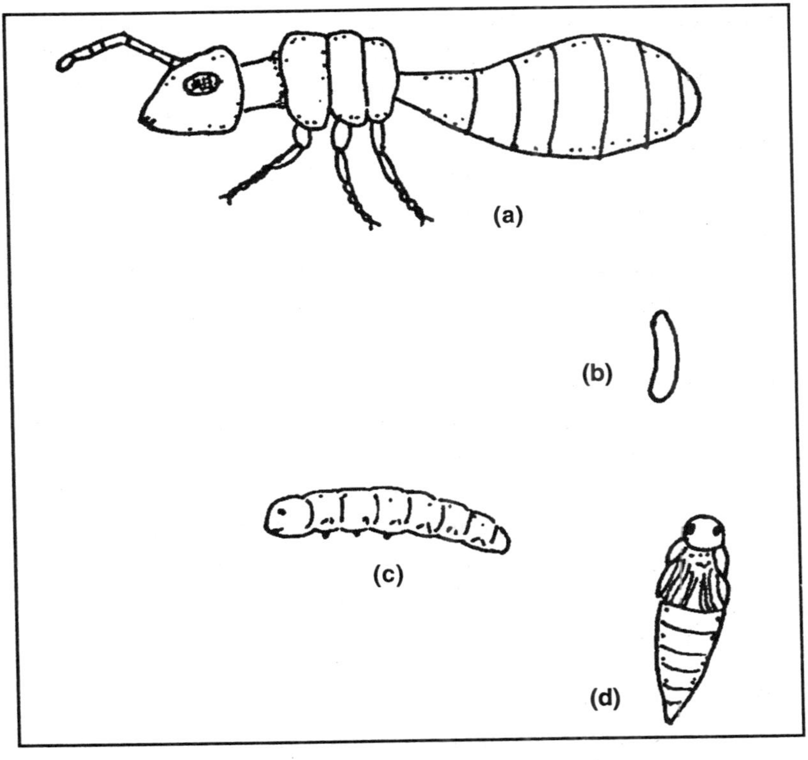

Figure 106: Life Cycle–Ant
(a) Adult; (b) Eggs; (c) Grub; (d) Pupa

grayish initially but turn brown at the time of maturation of this stage. The pupae are exarate type.

Life Cycle (Figures 106a–d)

At the beginning of monsoon season, when first rain appears the winged sexual forms of ants are emerged in large numbers. They mate during nuptial flight and females start egg laying while, males die after a short period of time after mating. After nuptial flight and end of mating, female settles on the land and findout suitable place for a chamber or nest for egg laying. She prepared a sealed chamber for herself and start egg laying in it. The eggs are typically hymenopteriform. Eggs hatch within a week or so. The newly emerged grubs are fed by the queen with the mouth secretions. The first generation is governed by the queen but lateron, when first generation is released, the workers perform tasks assigned to them by queen. Queen is covering authority of the colony. Thus, there is, caste system, division of labour, social behaviour and polymorphism in the ant.

Damage

1. Ants cause damage by biting to humans and cattles and thus causing pains.
2. The ants like sweet material hence, they set corowed, feed upon the sweet/food and spoil the food stuffs.
3. Winged forms sting to man, causing irritation and itchy lesions.
4. Contaminate food.

Control Measures

1. Ant nests should be destroyed and queen is killed.
2. Inject the petrol or kerosene or 10 per cent BHC or 5 per cent chlordane into the nests of ants by drilling the soil at several places of the nest.
3. Use baits like ant wells, ant tapes made up of lamp-wick soaked in a saturated solution of corrosive sublimate dissolved in ethyl alcohol and orange shellac.
4. Maintain cleanliness in the kitchen and house.

9. Wasps

Hymenoptera
Vespidae

(*i*) *Polistes herbraeus* (Fabr.)

Distribution

It is widely distributed in plains of India.

Marks of Identification

P. herbraeus is uniformly yellow coloured wasp with petiolate abdomen. It prepares simple and rounded nest in protected places on wall/ceilings/hollow trees, etc. It is smaller than *Vespa orientalis*

Nature of Damage

The wasp sting to the human being and inflict pain and swelling on the body.

(*ii*) *Vespa orientalis* Linnaeus (Figure 107)

Distribution

Plains of India.

Marks of Identification

It is larger than *P. hebraeus*. It shows deep brownish body colouration and transverse yellow bands on abdomen. It has shiny appearance. As like *P. hebraeus* it also construct its nest in protected areas such as humen dwellings, hollows in side walls, etc. *V.orientalis* construct a series of irregular but simple nests.

Nature of Damage

The wasp sting to humans for severe pains and swelling.

(*iii*) *Vespa mangifica* Smith.

Distribution

North India.

Marks of Identification

It is largest wasps amongst the species mentioned above. It has dark brown body colouration and robust look. It also shows numerous hairs on its body. Thus, it has hairy look too. It construct its nest on the ground and is abundant in the higher altitutes in north India.

Figure 107: Wasp (*Vespa orientalis*)

Nature of Damage

It is nuisance for apiculture business. It is very potential predator of honey bees. Hence, cause greatest damage to beehives and bee keeping business. The wasp can sting to humans for painfull response.

Life Cycle

First generation of wasps come out in the spring in old or newly constructed nests. The nests are constructed by queen itself in all the species reported above. The nest contains cells for egg laying. Queen lay her eggs in cells. The eggs hatched in cell produce larvae. The larvae are fed and reared by the queen and first generation is produced. The first generation contains only workers. The workers, then take care of further generations with respect to rearing, feeding, protection, etc. The food used for rearing the broods is masticated animal tissues collected by predation or biting off bits from dead animals. Sugary food is also given to the broods during their development by wasps. Full grown larva pupate in egg cell only. Pupal cells are capped for maintaining proper temperature for development of adult. During a single year pests can complete several generations. However, during cold, all workers and males die but, female survives and thus race is maintained. The pest hibernates in egg stage and becomes active in warm seasons of the year.

Control Measures

1. Nests of wasps should be destroyed by burning at night along with stages of pests.
2. Use of sugar baits in houses is also very good control strategy of this pest. In the bait Trichloroform is used as poison.

10. Bed Bugs

Hemiptera
Cimicidae

(*i*) *Cimex hemipterus* (*rotundatus*) Fabricious

Distribution

Cosmopolitan in distribution. Specially abundant in Tropical countries.

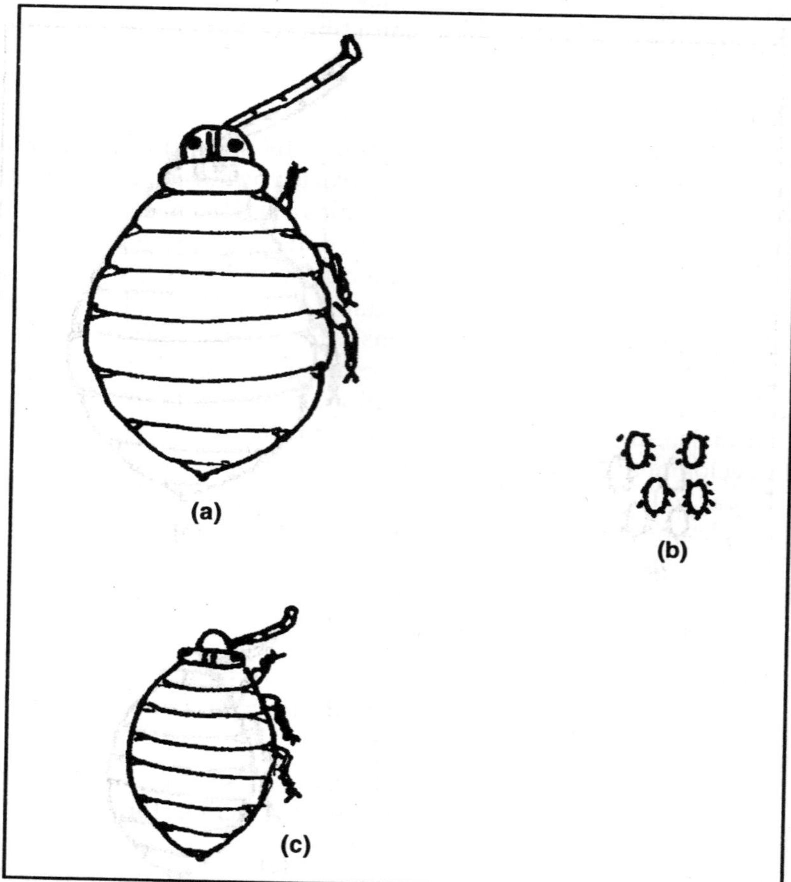

Figure 108: Bed Bug
(a) Adult; (b) Eggs; (c) Nymph

Marks of Identification

This species shows following characters:

1. Prothorax is less expanded laterally and extreme margins are not so flattened.
2. Colour is darker than *Cimex lectularis.*
3. Head is shorter and not broad.
4. Abdomen is not so orbicular.

Nature of Damage

It sucks the blood of man and poultry birds, chikens, bats, etc.

(*ii*) Common Bed Bug (*Cimex lectularis* Linnaeus)

Distribution

Cosmopolitan in distribution. It is abundant in temperate countries of the world.

Marks of Identification

1. It has mohogany red body colouration but lighter than *C. hemipterus*.
2. Head is broader than *C. hemipterus*
3. Extreme margins quite flattened.
4. Abdomen is orbicular.
5. Head is broder than *C. hemipterus*
6. Prothorax more expanded laterally.
7. Thoracic first segment is immediately behind the head.

Nature of Damage

Both nymphs and adults suck the blood of human being and cause itchy lesions. It attacks man, chikens, bats, domestic animals, etc. Bed bugs are most repulsive pests. The pest affects the fecundity of birds adversely in poultry business.

Life Cycle (Figures 108a–c)

Mated female lay her eggs in craks of furniture and walls, windows, racks, charpai, coats, cupboard, etc. Eggs are elongate whitish, slightly tapering at one end. A single mated female can lay about 80-500 eggs. 3-4 eggs are laid per female per day. Incubation period is 6-15 days depending upon climatic conditions. Hatchability of eggs is adversely affected after the treatment of DDT or BHC. Newly emerged nymphs are tiny, light, wingless, pear shaped, resembles with adult but are smaller in size. The nymphs moult for five times to become adults. The nymphal period is 5 to 6 weeks. In favorable condition, life cycle is completed within 6 weeks. Many over lapping generations (4) are completed during the year. The adult can survive for 1 year without food.

Control Measures

1. Treat the furnitures or beds or craks and crevices with malathion 1 per cent spray or crevices with malathion 1 per cent spray or Lindane 0.1 per cent spray.
2. Once the infestation gets a firm hold, it is very difficult to control this pest. Therefore, proper cleanliness within the house or human dwelling is necessary.

12. Rat Flea

Siphonaptera
Pulicidae
***Xenopsylla cheopis* (Rothschild)**

Distribution

Through out the world.

Marks of Identification

Rat flea adults are wingless and hard skinned insects. They show piercing and sucking type of mouth parts and concealed antennae. They have brownish black body colouration and large legs for jumping and have spiny appearance. Their larvae are whitish, legless, eyeless and very slender bodied. They have silky cocoons of small size and oval shape. Pupae are blackish brown and exarate type. Eggs are ovoid and relatively large.

Life Cycle (Figures 109a–d)

Mated female lay her eggs in dust or dirt or in host bedding or on the body of the host animal. Such eggs drop down on the ground. They hatch within 2-14 days depending upon climatic conditions. A single female can lay several hundred eggs. Newly emerged larvae feed on excreta of adult fleas or host animals such as mice, rat, Locusts

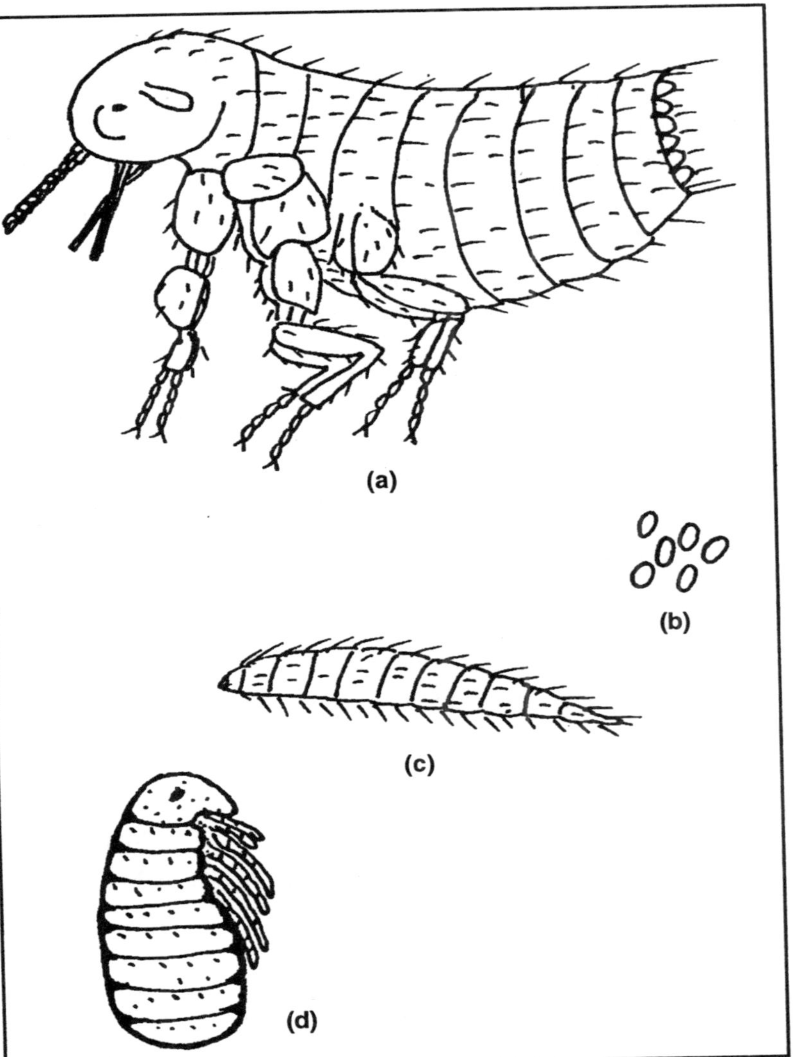

Figure 109: Rat Flea
(a) Adult; (b) Eggs; (c) Larva, (d) Pupa

and other rodents. The larva is with chewing type of mouth parts. The larval period is 1-5 weeks and pupal period is also 1-5 weeks. The full grown larva pupate in decaying organic matter or in soil in small, oval, silken cocoon. Winter is passed in the pupal stage. In favorable condition, the pest completes its life cycle within two weeks and extends for two years in adverse conditions. Many generations are completed by this pest during a year if conditions are favorable.

Nature of Damage

Rat flea acts as vector for "Bubonic plague" of human being which was responsible for wiping out the human population in early days. Rat flea bites with their piercing and sucking type of mouth parts for sucking the blood of man and cause irritating and itching skin lesions. Rat flea also acts as vector for transmitting murine endemic typhus and intermediate hosts of dog tapeworm and rodent tapeworm.

Control

1. Rat flea can breed on wild rodents. Therefore, periodic survey of rat flea on wild mammals and locusts is essential.
2. Human dwellings/houses should be clean and rat free and should well ventilated.
3. Houses are treated with lindane 1 per cent or malathion 0.5 per cent or chlordane 2 per cent spray.

13. House Fly

Diptera
Muscidae
***Musca domestica* Linnaeus**

Distribution

Cosmopolitan in distribution.

It is abundant in temperate region. In India, it is abundant in temperate region. In India, it is widely reported from Himalayan region.

Marks of Identification

Two winged housefly measures about 6-7 mm in body length. It is greyish with 4 whitish, longitudinal stripes on thorax. Antenna contain plumose arista. Hind wings are absent but tubular halters are present (a pair) on metathorax. *M. domestica* is identified by the feature, propleural depression setulose.

Musca nebulo Wiedemann

Distribution

Cosmopolitan

It is tropical species found in houses.

Marks of Identification

Propleural depression is not setulose, other characters are more or less similar to *M. domestica.* Maggot is opaque yellow, legless, tapering anteriorly and broder posteriorly; pupa is brown and barrel shaped. Eggs are typically elongated, curved and rounded at both the ends.

Life Cycles (Figures 110a–d)

A single mated female can lay about 500 eggs. Eggs are laid in decaying organic matter or cowdung, faeces, or any other filth. Eggs hatch within 1 to 12 days depending on climatic conditions. The larvae are apodous; first instars are active, may come out to the margin of the breeding material. The larvae moults for 3 times. The larval period is 5-8 days. The full grown maggots pupate in soil or dry decaying organic matter in barrel shaped brownish pupae. Pupal period is 4 to 6 days. Thus, life cycle of house fly is completed within 10 days in favorable conditions. Many generations are completed by housefly in a single year. Adults can survive maximum for 30 days.

Nature of Damage

Houseflies are amongst four 'F' s in epidemiology for spreading various diseases. The housefly acts a vector for transmitting following diseases.

1. Typhoid
2. Dysentry (both amoebic and bacterial)
3. Cholera
4. Yaws
5. T. B.
6. Leprosy
7. Trachoma

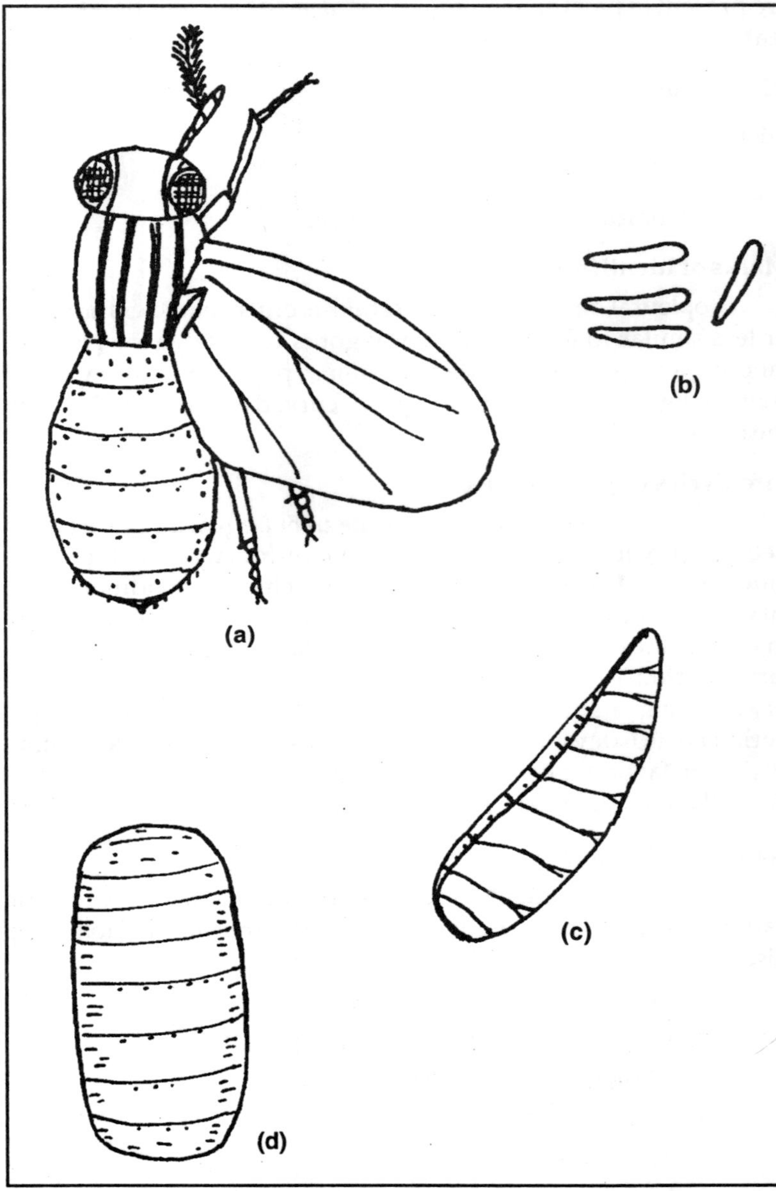

Figure 110a-d: Life Cycle of House Fly
(a) Adult; (b) Eggs; (c) Maggot; (d) Pupa

8. Anthrax
9. Gonorrhoea
10. Many helminthic diseases.
11. Viral diseases (Poliomyelitis)

Control Measures

1. Quick disposal of manure, garbage/etc.
2. Keeping premise clean.
3. Stopping outdoor/open toilets
4. Treating breeding places of housefly with any of the following–$CuSO_4$, formaldehyde, crude oil, lime.
5. Treating houses with baygon/malathion 2 per cent.

14. Sandfly

Diptera
Psychodidae
***Phlebotomus argentipes* Brunetti**

Distribution

Cosmopolitan in distribution.

In India sandfly is widely distributed in various states.

Marks of Identification

Sandfly measures about 2 mm in body length. It has small moth like appearance and dark body colouration. Hairs are arranged on legs, body and wings. Hairs resembles with lepidopterous scales hence, sandfly look like small moth. Its larvae are slender bodied and tapering toward both the ends. They bear spines arranged in transverse band on each segment. Its last segment contain too large bristles.

Life Cycles (Figures 111a–d)

A mated female lay eggs on decaying organic matter in damp or dark places. Eggs hatch within one week. The larvae feed on decaying organic matter and develop to its maturity within four instars. The larval period is one month. Full grown larva pupates in the soil/dry organic matter. Pupal period is 10 days. Thus, life cycle completes

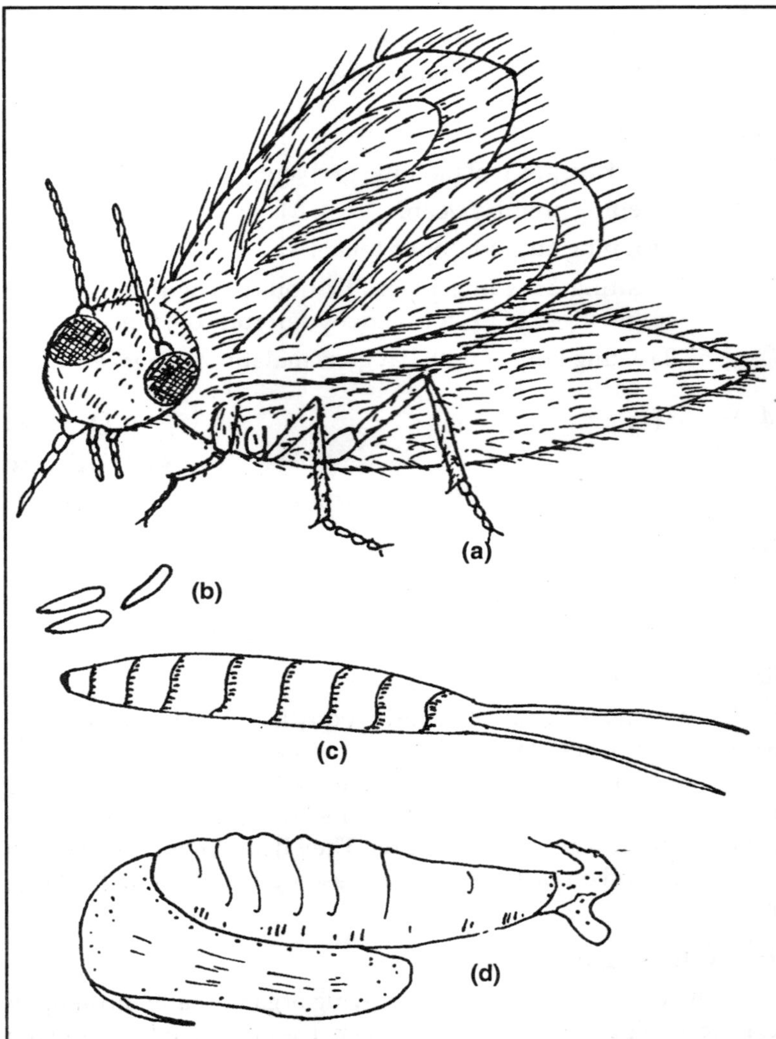

Figure 111a-d: Life Cycle of Sandfly
(a) Adult; (b) Eggs; (c) Maggot; (d) Pupa

within 6 weeks to 7 weeks. Many generations are possible in a single year if favorable conditions prevail.

Nature of Damage

Sandfly bites at night by piercing and sucking type of mouth parts. They suck the blood of humans and domestic animals. They prefer ankles, wrists, inner elbows, knee-joints, dorsum of feet, etc. for biting purpose. They cause intense itching and wheel like swelling. Sandfly acts as vector for transmitting Kala Azar, three day fever, tropical ulcers, etc. to man and "Leishmaniasis" and "harara" in domestic animals.

Control Measures

1. Treating the breeding places of sandflies with convential pesticides–such as malathion 0.2 per cent or Baygon 2 per cent
2. Keeping premises clean from decaying organic matter.
3. Sandfly bites can be protected at night by using mosquito nets at sleep.

15. Mosquitoes

Diptera
Culicidae

(*i*) *Anopheles* spp. (Figure 112)

Distribution

World wide in distribution. The genus *Anopheles* contain 500 species described in the world.

Marks of Identification

Mosquitoes are slender bodied insects with long legs. Maxillary palpi are equal to proboscis. Palpi of males is clubbed, scutellum are bar-shaped. Abdomen generally without scales or with few scales. Larva is without siphun tube but with two pairs of anterior clypeal hairs. Eggs are boat shaped and laid in fresh water. Body remains angular to sitting surface.

Distribution

Cosmopolitan

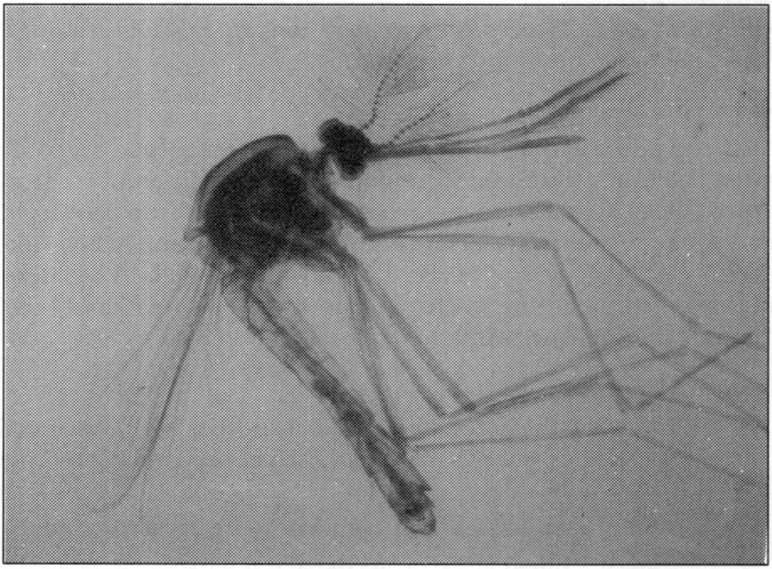

Figure 112: Mosquito–*Anopheles* sp.

(*ii*) *Culex* spp.

Diptera
Culicidae
***Culex* spp. (Figure 113)**

Distribution

Cosmopolitan.

Marks of Identification

Adult mosquitoes of culex species have unspotted wings. Palpi of female always shorter than proboscis. Palpi of male not clubbed and scutelluym is trilobed. Larva is without palmate hairs but with conspicuous siphun tube. Eggs are ovate and laid in dirty water near human dwellings.

(*iii*) *Aedes* spp. (Figure 114)

Adult mosquitoes have white bands on abdomen and on legs. Body colour is blackish brown with white bands on thorax and abdomen; maxillary palpi equal to proboscis. Eggs are elliptical. *Aedes* species are freshwater breeders.

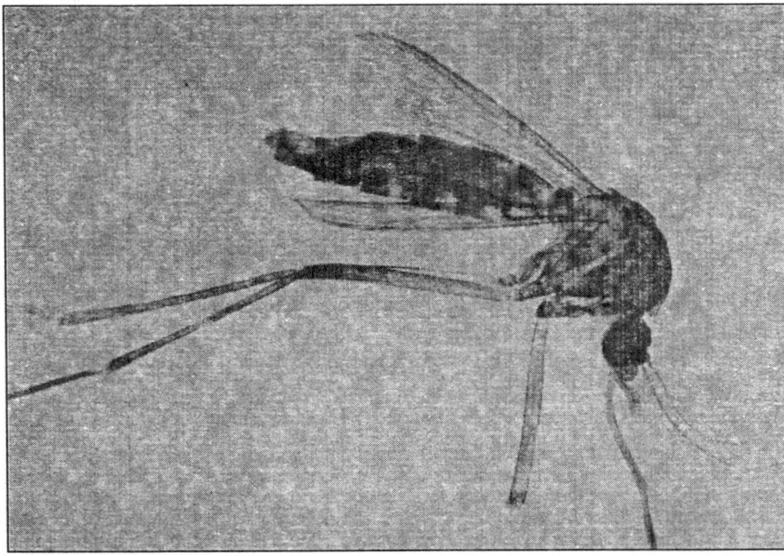

Figure 113: Mosquito–*Culex* sp.

Nature of Damage

Anopheles spp.

Ten species in adult stage acts as vector for transmitting pathogens for causing malaria in human being.

Malerial pathogens are:

1. *Plasmodium vivax*
2. *Plasmodium falsiparum*
3. *Plasmodium ovale*
4. *Plasmodium maleri*

Culex spp.

Culex pipiens fatigans acts as vector for transmitting nematode pathogen, *Wruchereria bancrofti* which cause Filaria to humans. *Culex* mosquitoes also acts as vector for transmitting a kind of virus for chikunguinea to man. These species are also responsible for Japnese encephlitis disease.

Aedes spp.

Aedes aegypti acts as vector for transmitting flavi virus which cause Dengue to human being and also chikun virus for causing chikungunia in human being.

Life Cycle

Mated females of mosquitoes lay their eggs in specific habitats as mentioned above. The eggs are hatched within 2-3 days. The newly emerged larvae are wingless which develop in aquatic medium and full grown as fourth instar within 2 days to 2 weeks. Pupae are also wrigglers and found in aquatic habitat. The pupal period is 2-3 days. Thus, within 6 days life cycle of mosquito is completed. During a single year many generations are completed by the mosquitoes.

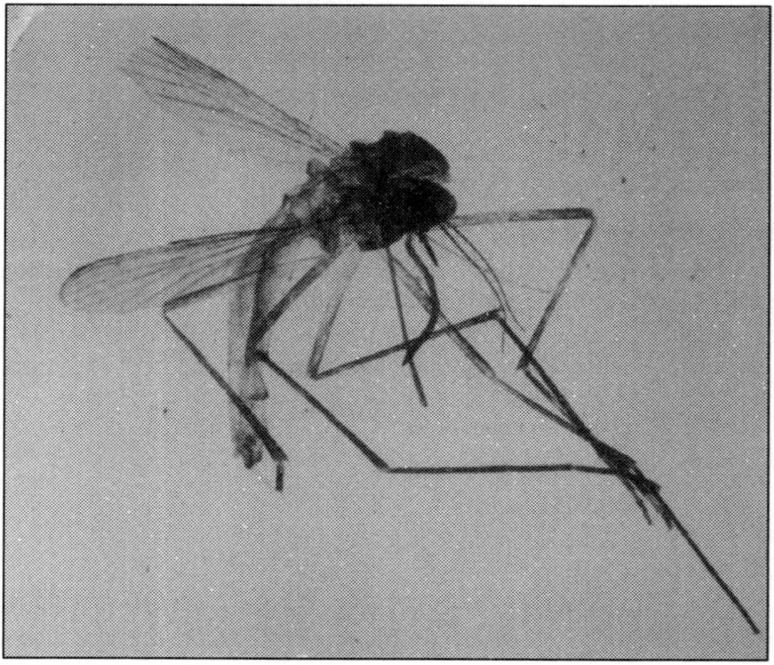

Figure 114: Mosquito–*Aedes* sp.

Control Measures

1. Disposal of breeding places of mosquitoes.
2. Spraying the houses with malathion 0.03 per cent or Bagon 2 per cent or chlordane 2 per cent will give good control of mosquitoes.
3. Fumigating breeding places of mosquitoes and houses with methylene dibromide.
4. Mosquitoes can also be controlled by biological means, using mosquitovorous fishes such as *Gambusia affinis* and *Poecilias reticulata* (Guppy fishes).
5. Mosquito nets are useful to get rid from mosquitoes at night sleep.

16. Eye Flies

Diptera
Chloropidae
***Siphunculina funicola* Meijere**

Distribution

It is widely distributed in India. Specially it has been reported from East and South India, Maharashtra, etc.

Marks of Identification

Adult fly is shining black and very small measuring about 2 mm in body length. It has 3 pairs of yellowish legs and one pair of transparent wings. It also shows a pair of halters on metathorax. Larvae are yellowish opaque.

Life Cycle

The mated female lay her eggs in batchs of 5-10 on decaying vegetables or decaying grasses bitten by heavy rain. They hatch within 2 days. Newly emerged larvae feed on decaying vegetables and grasses. The larva moult for 3 times within 5 days and then pupate in soil. Pupal period is 3 days. Thus, life cycle is completed within 10-21 days depending upon favorable/non favourable conditions. However, the pest occur in May and June.

Nature of Damage

Eye fly hovers infront of the eyes and produce typical noise. The flies feeds on the secretions of the eyes and secretions acts as transmitter of eye diseases, Opthalmia, conjunctivitis, etc.

Control Measures

1. Decaying grasses and vegetables should be immediately disposed.
2. Vegetable wastes should not thrown in open environment when there is out break of this fly.
3. Spray the decaying organic matter or grasses or vegetables thrown open, with pesticides, malathion 0.03 per cent or chlordane 0.2 per cent.
4. Keeping good hygienic and sanitary conditions of living environment will be good preventive control measure.

17. Tritoma Bug (Figure 115)

Hemiptera
Reduviidae
Tritominae

(*i*) *Tritoma rubrofaciata* sp.

Distribution

Widely distributed in central and south America and throughout the world.

Marks of Identification

The adult bug is with elongated and subcylindrical head, upper surface of head is without a buccal groove, rostrum 3 segmented and straight, antenna 4 segmented with apical portion narrow than other segments.

(*ii*) *T. rubida uhleri*

It shows long legs, and prompt biting action. It can fly at day time, 15 feet above ground level.

(*iii*) *T. protracta*

It is sluggish species, does not fly readily and bites hesitantly.

(*iv*) *T. diamidiata*

It fly at night about 7.00 pm to 11.00 pm.It can fly about 45 feet high from ground level and travel for considerable distance.

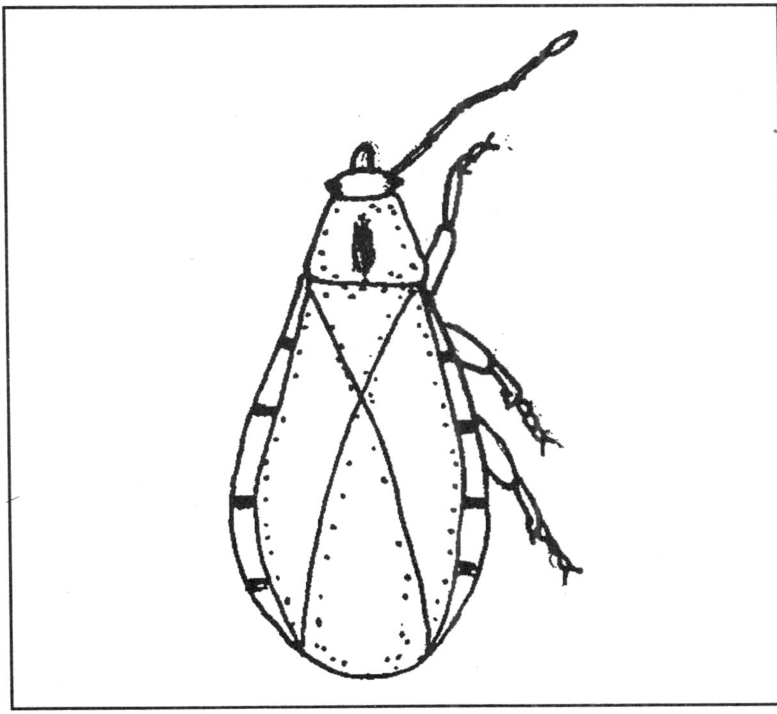

Figure 115: Adult–*Tritoma* sp.

Nature of Damage

The bugs bites human being and cause "Chagas disease" specially in western countries like USA and others.

Life Cycle

Preoviposition period in the bug is 9-28 days. A single mated female can lay about 100 to 700 eggs either singly or in groups or in masses. In a mass eggs are fastened together with the secretion secreted by the female. Incubation period is 7-60 days depending upon climatic conditions. Newly emerged nymphs are tiny with smaller eyes than adults. The nymph moults for five times and become adult. The nymphal period is 120 days to 1 year or more than even two year in adverse conditions.

Control Measures

1. Natural Control
   (*i*) Use of *Teleonomus farlai* (parasitoid–Hymenoptera)
   (*ii*) Use of *Solenopsis* sp. (Predatory insect)
   (*iii*) Ten mite species feed on this pest.
2. Use paints to wood material used in preparation of house.
3. Use light traps for collection and destruction of the pest.
4. Spray the houses with 2.5 per cent dieldrin.
5. Construction of wooden houses should be 6 to 10 feet above ground level.

18. Body Louse

Anopleura
Pediculidae

(*i*) *Pediculus humanus corporis* L.

Distribution

Cosmopolitan.

Marks of Identification

It shows greyish body colouration, diamond shaped head, elongate and wider thorax. Eggs deposited on cloths.

(*ii*) *Pediculus humanus capitis*

Distribution

Cosmopolitan.

Marks of Identification

Body colouration is greyish white with dark margins. Eggs deposited on hairs. It is associated with head.

(*iii*) Crab Louse (*Pthirus pubis*)

Distribution

Worldwide.

Marks of Identification

Body shape is oblong. Head is rectangular, abdomen is short, body segments indistinct. Claw size large and heavy. It is associated with pubic region of human being.

Nature of Damage

1. Both, nymphs and adults bite and cause cutaneous lesions, cause pediculosis, skin ulceration, etc.
2. Acts as vector for the diseases, typhus, trench fever, ERF, etc.

Life Cycle

Female louse lays about 100 eggs on body cloths/hairs or head/ pubic region as per the kind of louse. Eggs hatch within 7 days. There are 3 nymphal stages. Nymphal period is about 28 days. Thus, within 35 days life cycle is completed. Many generations are possible in a year.

Control Measures

1. Personal cleanliness is essential.
2. Lice are conrolled treating body, head and pubic region with " Lycil oil ".

Bibliography

Allard, H. A. (1910). Some New England Orthoptera observed in late October. *Ent. News*, 21(8), 352-7.

Allard, H. A. (1929). Physiological differentiation in overwintering individual of certain musical Orthoptera. *Can. Ent;* 61(9), 195-8, (W.L.13141).

Alexander, R. D. (1957). The taxonomy of the field crickets of the Eastern United States (Orthoptera: Gryllidae: *Acheta*). *Ann. ent. Soc. Am;* 50, 584-602.

Alexander, R. D. and Otte, D. (1967). The evolution genitalia and mating behaviour in crickets (Gryllidae) and other Orthoptera. *Misc. Publ. Mus. Zool. Univ. Mich.*, 133, 1-62.

Annandale, N. (1906). Notes on the fauna of a desert tract in southern India, Part II. Insects and Arachnida. *Mem. As. Soc.* Bengal, 1, 203-219.

Awate, M. R. and Sathe, T. V. (2007). On a new species of the genus *Gryllotalpa* Latreille (Orthoptera: Gryllotalpidae) from India. *Indian J. Environ. and Ecoplan;* 14 (3), 625-628.

Awate, M. R. and Sathe, T.V. (2008). On a new species of the genus *Gryllotalpa* Latreille (Orthoptera: Gryllotalpidae) from India. *J. Curr. Sci;* 12 (1), (In Press).

Banerjee, S. N. and Chatterjee, B. K. (1955). Studies on *Gryllotalpa africana* Palisot de Beauvis, a new major pest of the potato crop *Solanum tuberosum* Linn. and its control. *Indian. J. Ent.*, 17, 219-229, (W.L.22997).

Berthold, A. A. (1827). Latreille's naturliche Familieon des Tharreichs. Wiemar, Industr. Compt; 602.

Bey-Bienko, G. Y. (1966). The Orthopteran insects from Komodo and adjacent islands in Indonesia. *Zool. Zh;* 45, 1779-1796.

Bey-Bienko, G. Y. (1968). Orthopteroid insects of East Nepal. *Ent. Obozr;* 47, 106-130.

Bhargava, R. N. (1976). New record of *Loxoblemmus macrocephalus* Chopard (Orthoptera: Gryllidae) from Uttar Pradesh, N.W. India. *Newsl. Zool. Surv. India,* 2 (4), 141.

Bhargava, R. N. (1981). A new species of *Scleropterus* (Hann) (Insecta: Grylloidea) from N.W. India. *Geobios*, 8, 57-59.

Bhowmik, H. K. (1967a). Six new species of Gryllidae from India. *Proc.* 55th *Ind. Sc. Cong;* Part 3, 491-492.

Bhowmik, H. K. (1967b). Description of two new species of Gryllidae (Orthoptera: Insecta) from India. *Sci. and Cult;* 33, 244-246.

Bhowmik, H. K. (1967c). Five new species of crickets of the genus *Scapsipedus* Saussure (Insecta: Orthoptera) from India. *Proc. Zool. Soc.*, Calcutta, 20, 123-127.

Bhowmik, H. K. (1968). Two new species of Grylloidea (Insecta: Orthoptera) from India. *Proc. Zool. Soc.*, Calcutta, 21, 149-152.

Bhowmik, H. K. (1969). Studies on Indian Crickets (Orthoptera: Insecta), Part 1. *Zool. Anz., Bd;* 2, 143-152.

Bhowmik, H. K. (1970). The Gryllid fauna (Orthoptera: Insecta) of the Great Nicobar Island, India. *J. Zool. Soc.* India, 22 (1 and 2), 69-86.

Bhowmik, H. K. (1971). Two new species of Grylloidea (Orthoptera) from India. *Oriental Ins;* 5 (4), 503-506.

Bhowmik, H. K. (1975). Studies on Indian Crickets (Orthoptera: Insecta), Part III, *J. Bombay nat. Hist. Soc;* 72 (2), 368-382.

Bhowmik, H. K. (1976). On the Gryllid Fauna (Gryllidae: Orthoptera) of the Districts of the North Bengal with description of three new species. *Ind. Mus. Bull;* 11(1), 42-48.

Bhowmik, H. K. (1977a). On the new record of the Genus *Tartarogryllus* Tarbinskii and the description of *Velarifictorus andamanensis* (B.H.) from India. *Rec. Zool. Surv.* India,72, 363-366.

Bhowmik, H. K. (1977b). Studies on Indian Crickets (Orthoptera: Insecta) with description of two species. *Rec. Zool. Surv.* India, 73, 23-39.

Bhowmik, H. K. (1977c). Studies on some Indian Crickets with new distributional records of the subfamily Gryllinae (Gryllidae: Orthoptera). *Rec. Zool. Surv.* India, 73, 229-238.

Bhowmik, H. K. (1985). A checklist of the Gryllidae (Orthoptera) with Interterritorial distribution from the Eastern Himalayas. *Bull. Zool. Surv.* India, 7, (2-3), 185-193.

Bhowmik, H. K. and Halder, P. (1976). Some new records of Orthopteran Fauna from Parganas, West Bengal (India). *Newsl. Zool. Surv.* India, 2 (4), 143-144.

Biswas, S., Lahiri, A. R., Vasanth, M. and Ghosh, A. K. (1975). A preliminary study on the Insect fauna of Meghalaya. (Orthoptera: Gryllidae) Fifteen new records and other species. *Proc. Zool. Soc.* Calcutta, 28 (1), 23-26.

Blatchley, W. S. (1903). The Orthoptera of Indiana. 27th Ann. Rept. Ind. Dept. Geol. And Nat. Resources: 123-471.

Boake, C. R. (1983). Mating systems and signals in crickets. In: Orthopteran mating systems sexual compition in a diverse group of Insects (Ed. By D.T. Gwynne and G. K. Morris), 28-44. Boulder, Colorado, Westview Press.

Bolivar, I. (1890). Orthopteres de Africa de Museo de Lisboa. *Jorn. Sci.* Lisboa, 2 (1), 150-232.

Bolivar, I. (1900). Les Orthopteres de St. Joseph's College a Trichinopoly, 2e Partie. *Ann. Soc. Ent. Fr;* 68, 761-810.

Bolivar, I. (1901). Zoologische Ergebniss de dritten Asiatischen Forschungsreise des Grafen Eugen Zichy. Orthoptera, 223-243.

Bolivar, I. (1912). Orthoptera Acridiidae, Phasgonuridae, Gryllidae of the Seychelles. *Trans. Linn. Soc.* London, 15, 263-292.

Bolivar, I. (1922). Voyage de M. le baron Maurice Rothschild en Ethiopie et en Afrique Orientale anglaise (1904-1905).

Resultants Scientifique, Animaux articules. Orthopteres, Paris, Imprimere Nationale, 169-219.

Boucias, D. G., Maruniak, T. E. and Pendland, J. C. (1987). Characterization of aniridovirus isolated from the Southern mole cricket, *Scapteriscus vicinus*. *J. Invert. Pathol*; 50, 238-245.

Borror, D. J., Triplehorn, C. A. and Johnson, N. F. (1982). An introduction to the study of insect (sixth ed.). Published by Saunders College Publishing, Harcourt Brace College Publishers, Fort Wort Philadephia San Diego New York. Printed in the United States of America, 190.

Brunner, V. and Wattenwyl, C. (1873). Disposition Gryllodeorum. *Mitth. Sch. Weiz. Ent. Ges*; 4,164-170.

Brunner, V. and Wattenwyl, C. (1893). Revesion du systeme des Orthopteres et description des especes rapportees par M. Leonardo fea de Birmanie *Ann. Mus. Civ. St. nat. Genova*, 13, 1-230.

Burmeister, H. C. (1838). Handbuch der Entomologie. Zweiter Band, zweite Abtheilung, Kaukerke,Gymnognatha. Berlin, Th. Ch. F. Enslin, 397-756.

Cade, W. H. (1980). Alternative male reproductive behaviors. *Florida Entomologist*, 63, 30-45.

Cade, W. H. and Cade, E. S. (1992). Male mating success, calling and searching behaviour at high and low densities in the field cricket *Gryllus integer*. *Anim. Behav*; 43, 49-56.

Castner, J. L. (1987). *Larra bicolor* a natural enemy of mole crickets. *Hort. Sci.*, 22, 1060.

Castner, J. L. (1988). Biology of the mole cricket parasitoid *Larra bicolor* (Hymenoptera: Sphecidae), 423-432.

Caudell, A. N. (1927). On a collection of Orthopteroid insects from Jawa made by owen Bryant and William Palmer. *Proc. U.S. Nat. Mus*; 71, 1-42.

Chatterjee, R. N. (1955). Ecology and control of cricket *Gymnogryllus humeralis* Walker (Orthoptera, Gryllidae), a minor pest of *Casuarina equisetifolia* Forst. seedling in Bombay State, with a list of other pests of *Casuarina*. *Ind. Forest. Rec.*, 81, 509-511.

Chopard, L. (1918). Etude des Gryllides du Museo civico di Storia naturale di Genova. *Ann. Soc. Ent. Fr*; 86, 509-574.

Chopard, L. (1924a). Note sur quelques orthopteres cavernicoles de Sumatra et de Jawa. *Ann. Soc. Ent. Fr; 93,* 189-192.

Chopard, L. (1924b). On some cavernicolous Orthoptera and Dermaptera from Assam and Burma. *Rec. Ind. Mus;* 26, 81-92.

Chopard, L. (1924c). Note sur quelques Orthopteres cavernicoles de Sumatra et de Java. *Ann. Soc. Ent. Fr;* 93, 189-192.

Chopard, L. (1925a). Descriptions de Gryllides nouveaux de Java. *Treubia,* 6, 151.

Chopard, L. (1925b). Results of Dr. E. Mjoberg's Swedish Sci. Expedition to Australia (1910-1913). *Ark. Zool;* 18 A (6), 1-57.

Chopard, L. (1925c). Descriptions de *Gryllides nouveaux* (Orthopteres). *Ann. Soc. Ent. Fr;* 94, 291-332.

Chopard, L. (1925d). The Gryllidae of Ceylon in the Britsh Museum Collections. *Ann. Mag. nat. Hist;* (9), 15, 505-536.

Chopard, L. (1927). Descriptions de *Gryllides nouveaux. Ann. Soc. Ent. Fr;* 96, 147-174.

Chopard, L. (1928a). Revision of the Indian Gryllidae. *Rec. Ind. Mus;* 30, 1-36.

Chopard, L. (1928b). Additional notes on the Gryllidae of Ceylon. *Spol. Zeyl;* 14, 197-208.

Chopard, L. (1929a). Fauna of the Batu Cave, Selangor. 12. Orthoptera and Dermaptera. *Jour. Fed. Malay.States. Mus;* 14, 366-371.

Chopard, L. (1930). The Gryllidae of Sarawak. *Sarawak Mus. J;* 4, 1-42.

Chopard, L. (1931*a*). On Gryllidae from the Malay Peninsula. *Bull. Raffles Mus;* 6, 124-149.

Chopard, L. (1931*b*). Dr. E. Mjoberg's Zoological Collections from Sumatra. 12. Gryllidae *Ark. J. Zool;* 23A, (9), 1-17.

Chopard, L. (1933*a*). New species of Indian Gryllidae (Orth.). *Stylops,* 11, 115–120.

Chopard, L. (1933*b*). Gryllidae de 1'Inde meridionale *Rev. Suisse Zool;* 40, 161-167.

Chopard, L. (1935). On a collection of Gryllidae from India made by Mr. B.M. Fletcher.*Ann. Mag. Nat. His;* (10), 16, 284–296.

Chopard, L. (1936). The Tridactylidae and Gryllidae of Ceylon. *Ceylon. J. Sc;* B. 20, 9-87.

Chopard, L. (1941). Results of the Oxford Univ. Exped. to Sarawak (Borneo), (1932). Gryllacridae and Gryllidae. *Ent. mon. Mag;*76, 189-224.

Chopard, L. (1951). A revision of the Australian Grylloidea. *Rec. S. Austl. Mus; Adelatide,* 9, 397-533.

Chopard, L. (1954). Le difficile probleme de systematique pose par les grillons. *Ann. Mus. Congo. Tervuren, Zool;* 326-327.

Chopard, L. (1961). Les divisions du genre *Gryllus* besees sur 1'etude de 1' appareil Copulateur (Orthoptera: Gryllidae). *Eos;* 37, 267-287.

Chopard, L. (1969). The Fauna of India and Adjacent countries. Orthoptera, 2, Grylloidea. The Manager of Publications, Govt. of India, Delhi, xviii, 421.

Chopard, L. (1970). Description de Gryllidae cavernicoles nouveaux (Orth. Phalangopsidae). *Bull. Soc. Ent. Fr;* 75, 7-123.

Chopard, L. and Beier, M. (1967). Orthopterorum Catalogus, Gryllides Pars 10, Uitgeverij Dr W. Junk's Gravenhage, 211, 1-21.

Chopard, L. and Beier, M. (1968). Orthopterorum Catalogus, Gryllides, Pars 12, Uitgeverij Dr W. Junk's-Gravenhage, 288, 213-500.

Chopard, L. and Chatterjee, N.C. (1937). Entomological investigations on the spike disease of Sandal. Dermaptera and Orthoptera. *Ind. For. Rec.,* 31, 3.

Chopard, L. and Dreux, P. (1966). Contribution a 1'etude des orthopteroides du Nepal. *Ann. Soc. Ent. Fr;* 2, 601-616.

David, M. J. (1994). Palynology and the Mole cricket. Canadian Museum of Nature Ottawa, Ontario, Canada. Elliot Burden, President's Message, CAP Newsletter. 17(1), 2-3.

De Geer, Carl. (1773). Memoires pour server a 1' istoire des Insects, Stockhlm, 3, 44, 696.

Fabricius, J. C. (1775). Systema Entomologicea systems Insectorum classes, ordines, genera, species, adjectis synonymis, locies. Flensburg et Leipzig, Korte, 832.

Fernando, W. (1957a). New species of Insects from Ceylon (1). *Ceylon J. Sci.*, I, 7-18.

Fernando, W. (1957*b*). New species of Insects from Ceylon (2). *Ceylon J. Sci.*, I, 73-76.

Fernando, W. (1957*c*). Contributions to a knowledge of the insects of Ceylon. 2. *Ectatoderus sandasagarai*. *Ceylon J. Sci;* (B) 25, 201.

Forrest, T. G. (1980). Phonotaxis in mole crickets: its reproductive significance. *Florida Entomologist;* 63, 45-53.

Forrest, T. G. (1983). Calling Songs and mate choice in mole crickets Gwynne, D.T., Morris, G. K. (eds.) Orthopteran mating systems: sexual competition in a diverse group of insects. Westview; Boulder, Co; 185-204.

Frank, J. H. (1990). Mole crickets and other arthropod pests of turf and pastures, 131-139 in Habeck D.H., Bennett F.D., Frank J.H. (eds.) Classical biological control in the Southern United States. Southern cooperative series Bulletin, 355, i-viii; 1-197.

Frank, J. H. (1994). Biological control of pest mole crickets, Rosen, D., Capinera, J.L., Bennett, F.D. (eds.) Biological control and IPM: The Florida Experience. Intercept and over, England, 343-352.

Frank J. H., Parkman J. P. and Bennett F. D. (1995). *Larra bicolor* (Hymenoptera Sphecidae), a biological control agent of *Scapteriscus* mole crickets (Orthoptera: Gryllotalpidae), established in northern Florida. *Florida Entomologist*, 78, 619-623.

Frank, J. H. and Parkman, J. P. (1999). Integrated pest management Review, 4, 39-52.

Frank, H. and Robert, Y. (1994). Hey ! the war isn't Over yet ! *Florida Turf Digest*, 21-24.

Fritz, G. N. (1983). A technique for separating mole crickets from soil. *Florida Entomologist*, 66, 360-362.

Fulton, B. B. (1930). A new species of *Nemobius* from North Carolina (Orthoptera: Gryllidae). *Ent. News*, 41(2), 38-42.

Fulton, B. B. (1952). Speciation in the field cricket. *Evolution,* 6(3), 283-95.

Ghosh, C. C. (1912). The big brown cricket *Brachytrypes achatinus* Stoll Calcutta. *Mem. Dept. Agric. Ind. Ent. Ser;* 4(3), 161-182.

Ghouri, A. S. K. and McFarlane, J. E. (1958). Occurrence of a macropterous from of *Gryllodes sigillatus* Walker (Orthoptera: Gryllidae) in laboratory culture. *Can.J. Zool;* 837-838.

Giebel, C. G. (1870). Description of *Gryllotalpa siamensis. Ztschr. f. die gesammt. Naturwiss;* 35, 48.

Guerin-Meneville, F. E. (1833). Insects, in Voyage aux Indes Orientales par C. Belanger. Zoologie, 441-512.

Guerin-Meneville, F. E. (1845-1850). Voyage en Abyssinie execute pendant les annees 1839-1843 (par M. Lefevre). Paris, Bertrand, 239-390.

Gupta, V. K. (1988). Advances in parasitic Hymenoptera Research. *Brill; Leiden,* 546.

Gupta, R. K. (2004). Advancements in insect biodiversity (Agrobios., Jodhpur), 1-324.

Haan, W. D. (1842). Bijdragen tot de Kentnis der Orthoptera, 1842-44, 45-248.

Hebard, M. (1926). Dermaptera and Orthoptera: In Insects of Hawaii, Johnston Island and Wake Island, by E.A. Bryan (Jr.) *et al. Bull. Bishop Mus;* 31, 82-88.

Hudson, W. G. (1994). Life cycles and population monitoring of pest mole crickets Leslie, A.R. (ed.) Integrated Pest Management for Turfgrass and Ornamentals. CRC Press; Boca Raton, FL, 345-349.

Hudson, W. G.; Frank, J. H. and Castner, J. L. (1988). Biological control of *Scapteriscus* spp. Mole crickets (Orthoptera: Gryllotalpidae) in Florida. *Bull. Ent. Soc. Am.,* 34,192-198.

Jacobson, G. G. and Bianchi, V. L. (1905). *Orthopteres et Nevropteres de V Empire russe et des regions voisines.* Saint-Petersbourg, 10, 952.

Johnstone, D. E. (1975). A gynandromorphy cricket, *Gryllus bimaculatus* (Grylloptera: Gryllidae). *Can. J. Zool;* 53, 1505-1513.

Jones, S. and Chopard, L. (1936). On an interesting new genus of Gryllid from India. *Ann. Mag. Nat. Hist*; 18 (10), 234-238.

Kaltenbach, A. V. (1979). Ergebnisse der osterreichisch indischen Andamanea. Expedition 1976. Saltatoria (Insecta: Orthopteroidea). *Ann. Naturhistor. Mus. Wien*; 82, 605-614.

Karny, H.(1909). Ost-afrikanische Orthopteren Sammelausbeute von A. Borgert (1904-1905). *Zool. Jahrb. Syst.*, 27., 477-480.

Khalifa, A. (1950). Sexual behaviour in *Gryllus domesticus* L. *Behaviour*, 2, 264-274.

Khan, N. H. (1954). Ecological observations on the eggs of *Gryllodes sigillatus* Wallker. *Indian. J. Ent.*, 16, 24-26.

Kirby, W. F. (1906). Orthoptera Saltatoria, Part I. (Achetidae et Phasgonuridae). A Synonymic Catalogue of Orthoptera, London 2, 562

Latif, A. and Asghar, M. A. (1957). Two new species of Gryllidae. *Proc. 8th Pak. Sci. Con*; 1956, *Sec. biol*; 69.

Latreille, P. A. (1804). Order Second. Orthopteres; Orthoptera in Histoire naturelle, generale et particuliere des Crustaces et Insectes. Paris 12, 81-164.

Latreille, P. A. (1829). Insectes dans Le Regne animal de Cuvier, 3, 1-329. Paris, Masson.

Le Guillou, E. J. F. (1841). Description des Orthopteres nouveaux recueillis pendant son voyage de circumnavigation sur la corvette *Zelee. Rev. Zool*; Paris, 4, 292-295.

Linnaeus, Carl Von. (1758). Systema Naturae per Regna tria Naturae, (Ed.10) Stockholm 1, 823.

Mark, J. F. and David, A. G. (2001). Divergence between the courtship songs of the field crickets *Gryllus texensis* and *Gryllus rubens* (Orthoptera, Gryllidae). *Ethology*, 107 (12), 1075-1085.

Mayr, E. (1969). Principles of Systematic Zoology. Tata McGraw Hill Publishing Co. Ltd; x + 428.

Meagher, R. L. and Frank, J. H. (1998). *Larra bicolor* (Hymenoptera: Sphecidae: Larrinae) collected in pheromone–and phenyl/ acetaldehyde-baited traps. *Florida Entomologist*, 81, 555-556.

Nguyen, K. B. and Smart, G. C. (1992a). Addendum to the morphology of *Steinernema scapterisci*. *J. Nemat;* 24, 478-481.

Nguyen, K. B. and Smart, G. C. (1992b). Life cycle of *Steinernema scapterisci* Nguyen and Smart. *J. Nemat;* 24, 160-169.

Omachi, F. (1926). Preliminary note on a gynandromorphy. *Madasumma marmorata*. *Proc. Imp. Acad;* Tokyo, 2, 554-558.

Otte, D. and Cade, W. (1984). African crickets (Gryllidae). 6. The genus *Gryllus* and some related genera (Gryllidae, Gryllinae). *Proc. Acad. Nat. Sci. Phila;* 136, 98-122.

Pallas, P. S. (1771). Reisen durch verschiedene Provinzen des Russischen in den Jahren 1768-1774. Petersburg Akadem. Buchhandl. 1: 468.

Pantel, J. (1896). Notes: Orthopterologiques. *Ann. Soc. Esp. Hist. Nat;* 25, 47-118.

Parkman, J. P. and Frank, J. H. (1993). Use of a sound trap to inoculate *Steinernema scapterisci* (Rhabditida: Steinernematidae) into pest mole cricket populations (Orthoptera: Gryllotalpidae). *Florida Entomologist*, 76, 75–82.

Parkman, J. P., Frank, J. H., Nguyen, K. and Smart, G. C. (1994). Inoculative release of *Steinernema scapterisci*. (Rhabditida: Steinernematidae) to suppress pest mole crickets (Orthoptera: Gryllotalpidae) of golf courses. *Environ. Entomol;* 23, 1331-1337.

Rambur, J. P. (1839). Fauna entomologiques de 1'Andalousie. II. Dermapteres et Orthopteres, 1-94.

Randell, R. L. (1964). The Male Genitalia in Gryllinae (Orthoptera: Gryllidae) and a Tribal Revision. *Can. Ent;* 96, 1585-1607. (W.L. 1314).

Rao, A. S. and Mani, M. S. (1974). Ecology and Biogeography in India. Dr W. Junk b.v. Publishers. The Hague. XIX + 773.

Rehn, J. A. G. (1902). Notes on the Orthoptera of New Mexico and western Texas. *Proc. Acad. Nat. Sci. Phila;* 54, 726-727.

Rehn, J. A. G. (1903). Description of a new genus of Gryllidae with a note on the genus *Aphonogryllus*. *Ent. News*, 14, 258-260.

Rehn, J. A. G. and Hebard, M. (1915). The genus *Gryllus* (Orthoptera) as found in America. *Proc. Acad. Nat. Sci. Phila;* 67, 292-322.

Sailer, R. I., Reinert, J. A., Drion, B., Philip, B., Kepner, R. L., Forrest, T. G., Hudson, W. G. and Walker, T. J. (1984). Mole crickets in Florida. University of Florida, Gainesville, Institute of Food and Agricultural Sciences Bulletin, 846, 1-54.

Sandrasagara, T. R. (1954). Checklist of the Tridactylidae and Gryllidae (Insecta, Orthoptera) of Ceylon with records of distribution. *J. Bombay nat. Hist Soc;* 52, 540-562.

Sathe, T.V. and Bhoje, P.M. (2000). Biolgical Control of Insect Pests. DPH, Delhi, pp. 1-127.

Sathe, T.V., 2003. Agrochemicals and Pest Management. Daya Publishing House, New Delhi, p. 217

Sathe, T.V., 2005. Basic Entomology: A Practical Manual. Daya Publishing House, New Delhi, p. 118.

Sathe, T.V. & Margaj, G.S., 2001. Cotton Pests and Biocontrol Agents. Daya Publishing House, New Delhi, p. 166.

Sathe, T.V. & Shinde, K.P., 2008. Dragonflies and Pest Management. Daya Publishing House, New Delhi, p. 179.

Sathe, T.V., 2006. Fundamentals of Beekeeping. Daya Publishing House, New Delhi, p. 100.

Sathe, T.V. & Jadhav, B.V., 2008. Indian Pest Aphids. Daya Publishing House, New Delhi, p. 211.

Sathe, T.V. *et al.*, 2003. Indian Pest Parasitoids. Daya Publishing House, New Delhi, p. 145.

Sathe, T.V. & Bhosale, Y.A., 2001. Insect Pest Predators. Daya Publishing House, New Delhi, p. 169.

Patil, Vaishali J. & Sathe, T.V., 2003. Insect Predators and Pest Management. Daya Publishing House, New Delhi, p. 216.

Sathe, T.V. & Girhe, B.E., 2002. Mosquitoes and Diseases. Daya Publishing House, New Delhi, p. 96.

Sathe, T.V. *et al.*, 2005. Practicals in Basic Entomology. Daya Publishing House, New Delhi, p. 84.

Sathe, T.V. & Jadhav, A.D., 2001. Sericulture and Pest Management. Daya Publishing House, New Delhi, p. 197.

Sathe, T.V. & Thite, S.H., 2004. Shoot Feeding and Sericultural Trends. Daya Publishing House, New Delhi, p. 157.

Sathe, T.V., 2009. Textbook of Forest Entomology. Daya Publishing House, New Delhi, p. 234.

Sathe, T.V., 2004. Vermiculture and Organic Farming. Daya Publishing House, New Delhi, p. 122.

Saussure, H. D. (1874). Etudes sur les insects Orthopteres. *Miss. Sci. Mex. Amer. Centr. (Rech. Zool; 6, 1)* 3, 293-516.

Saussure, H. D. (1877). Melanges Orthopterologiques. Ve fascicule. Gryllidae. *Mem. Soc. Geneve*, 25, 1-352.

Saussure, H. D. (1878). Melanges Orthopterologiques. VIe fascicule. Gryllidaes. *Mem. Soc. Geneve*, 25, 369-702.

Sassure, H. D. (1894-1897). Gryllidae. Biologia Centrali-Americana. *Insecta Orthoptera*, 1, 198-284.

Schimmer, F. (1911). *Mymecophila escherichi*, eine neue termitophile Ameisengrille, in Escherich Termitenleben aus Ceylon. *Iena, G. Fischer*, 233-235.

Scudder, S. H. (1862). Material for a monograph of the North American Orthoptera, including a catalogue of the known New England species. Bost. *Jour. Nat. Hist;* 7, 409-480.

Scudder, S. H. (1869). Revision of the large stylated fossorial crickets. *Mem. Peabody Academy of Sc.*, first Mem. Salem, 1-28.

Scudder, S. H. (1901). The species of *Gryllus* on the west coast. *Psyche*, 9 (307), 267-270.

Scudder, S. H. (1902).The species of *Gryllus* found in the United States east of the Sierra Nevadas. *Psyche*, 9 (309), 291-296.

Searcy, W. A. and Andersson, M. (1986). Sexual selection and the evolution of song *A. Rev. Ecol. Syst;* 17, 507-533.

Serville, M. A. (1831). Revue methodique des Insectes. Orthopteres. *Ann. Sc. Nat;* 22, 1-262.

Serville, M. A. (1839). Histoire naturelle des Insectes. Orthopteres. Paris, Roret, 776.

Shiraki, T. (1930). Orthoptera of the Japanese Empire. I. Gryllotalpidae and Gryllidae. *Insecta Matsumurana*, 4, 181-252.

Shishodia, M. S. and Tandon, S. K. (1975a). New record of *Ornebius pendleburyi* Chopard from India. *Newsl. Zool. Surv.* India, 1(1), 6.

Shishodia, M. S. and Tandon, S. K.(1975b). General Insect Survey of Tripura State. *Newsl. Zool. Surv.* India, 1(2), 14.

Shishodia, M. S. and Tandon, S. K. (1977). Some new records of Grylloidea (Insecta: Orthoptera) from Andaman Islands. *Newsl. Zool. Surv.* India, 3 (3), 125-126.

Short, D. E. and Koehler, P. G. (1979). A sampling technique for mole crickets and other pests in turfgrass and pasture. *Florida Entomologist*, 62, 282-283.

Shrivastava K. P. (1996). A Textbook of Applied Entomology. Kalyani publishers, Vol. II, 1-507.

Sinha, K. M. and Agrawal, S. M. (1977). A new cavernicolous Orthoptera, *Kempiola shankari* (Orthoptera: Phalangopsidae) from Madhya Pradesh. *Ind. For.*, 103(2), 150-152.

Smart, G. C. and Khuong, N. B. (1986). Biological control of mole crickets with a neoaplectanid nematode. Soil Crop. Sci. Soc. Florida, Proc; 45, 192.

Snodgrass, R. E. (1935). Principles of insect morphology Mc-Graw-Hill, New York. 1-667.

Snodgrass, R.E. (1937). The male genitalia of Orthopteroid insects. Smithson. *Misc.Collect*; 96, 1-107.

Stoll, C. (1813). Representation exactement coloree d'apres nature des Spectres ou Phasmes, des Mantes, des Sauterelles, des Grillons, des croquets et des Blattes qui se trouvent dans les quatre parties du Monde, I'Europe, I'Asie, I'Amerique. Amsterdam, J. C. Sepp et fils, un in 4, 180.

Suzuki, K. (1934). Gynandromorphs in *Homoeogryllus japanicus* de Hann. *Proc.* IMP. *Acad*; Tokyo, 10, 509-512.

Tandon, S. K. and Shishodia, M. S. (1972). Notes on the collection of Grylloidea (Orthoptera) from NEFA, India. *Oriental Ins*; 6(3), 281-292.

Tandon, S. K. and Shishodia, M. S. (1974). Descriptions of two new species of *Velarifictorus* (Orthoptera: Grylloidea) from India. *Oriental Ins*; 8(3), 299-302.

Tandon, S. K.; Shishodia, M. S. and Dey, A. (1976). On a collection of Orthoptera (Insecta) from the Kanha National Park, Mandla, Madhya Pradesh, India. *Newsl. Zool. Surv.* India, 2 (4), 167-170.

Townsend, B. C. (1980). A taxonomic study of two similar species of *Teleogryllus* (Orthoptera: Gryllidae), *J. nat. Hist;* 14,153-162.

Tweedie, M. (1965) Gynandromorphs-both male and female but not hermaphrodites. *Animals* (London), 8 (1), 22-24.

Uvarov, B. P. (1925). Orthoptera (except Blattidae) collected by Prof. Gregory' expedition to Yunnan. *J. Asiat. Soc. Bengal,* 20, 313-335.

Uvarov, B. P. (1935). A new name for the genus *Gryllus auct. Nec L. Ann. Mag. Nat. Hist;* 10 (20), 120.

Uvarov, B. P. (1940). Twenty-eight new generic names in Orthoptera *Ann. Mag. Nat. Hist;* 11 (5), 173-176.

Vasanth, M. (1978). Three unknown females of crickets (Orthoptera: Gryllidae) from Northeast India with notes on their distribution. *Entomon,* 3(2), 273-276, (W.L. 18270)

Vasanth, M. (1980). Two new species of crickets of the genus*Turanogryllus* Tarbinskii (Orthoptera: Gryllidae) from Maharashtra. *Biovigyanam,* 6, 115-119.

Vasanth, M. (1988). Observation on the mating behaviour of *Gryllodes sigillatus* Walker (Insecta: Orthoptera: Gryllidae) *Rec. zool. Surv.* India, 85(3), 411-418.

Vasanth, M. (1993). Studies on crickets (Insecta: Orthoptera: Gryllidae) of Northeast India. *Rec. Zool. Surv.* India, 132, 1-177.

Vasanth, M., Lahiri, A. R., Biswas, S. and Ghosh, A. K. (1975). Three new species of Gryllidae (Insecta: Orthoptera). *Oriental Ins;* 9 (2), 221-228.

Vickry, V. R. (1977). Taxon ranking in the Grylloidea and Gryllotalpidea. *Lyman ent. Mus. Res. Lab. Memoir,* 4, 32-43.

Walker, F. (1859). Characters of some apparently undescribed Ceylon Insects. *Ann. Mag. Nat. Hist;* 3 (4), 217-223.

Walker, F. (1869). Catalogue of the specimens of Dermaptera Saltatoria in the collection of the British Museum. Gryllidae, London, British Museum, 1-117.

Walker, F. (1871). Catalogue of the specimens of Dermaptera Saltatoria in the collection of the British Museum, Supplement, 1-18.

Walker, T. J. (1966). Annotated Checklist of Oecanthinae (Orthoptera: Gryllidae) of the World. *Florida Entomologist,* 49 (4), 265-277.

Walker, T. J. (1982). Sound traps for Sampling mole cricket flights (Orthoptera: Gryllotalpidae: Scapteriscus). *Florida Entomologist,* 65 (1), 105-110.

Walker, T. J. (1986). Monitoring the flights of Field crickets *Gryllus* spp. and a tachinid fly *Euphasiopteryx ochracea* in north Florida. *Florida Entomologist,* 69, 678-685.

Walker, T. J. (1989). A live trap for monitoring *Uphasiopteryx* and tests with *E. ochracea* (Diptera : Tachinidae). *Florida Entomologist,* 72, 314-319.

Walker, T. J. (1996). Acoustic methods of monitoring insect pests and their natural enemies. 245-247 In D. Rosen, J. L. Capinera, and F.D. Bennett (eds.), Pest management in the Subtropics: Integrated Pest Management-A Florida Perspective. Intercept; Andover, UK.

Walker, T. J. and Fritz, G. N. (1983). Migratory and local flights in mole cricket, *Scapteriscus* sp. (Gryllotalpidae). *Environ. Entomol;* 12, 953-958.

West, M. J. and Alexander, R. D. (1963). Sub-social behaviour in a burrowing cricket *Anurogryllus muticus* De Geer (Orthoptera: Gryllidae). *Ohio J. Sci;* (1), 19-24.

---

*W.L. is the World list number of the registered journal which is mentioned only for the first citation of the journal in the bibliography.

Index